MARINE BIOLOGY

ALGAE

ECOLOGY, ECONOMIC USES AND ENVIRONMENTAL IMPACT

Marine Biology

Additional books in this series can be found on Nova's website
under the Series tab.

Additional e-books in this series can be found on Nova's website
under the e-book tab.

Environmental Science, Engineering and Technology

Additional books in this series can be found on Nova's website
under the Series tab.

Additional e-books in this series can be found on Nova's website
under the e-book tab.

MARINE BIOLOGY

ALGAE

ECOLOGY, ECONOMIC USES AND ENVIRONMENTAL IMPACT

DAGMAR KRUEGER
AND
HELGA MEYER
EDITORS

Nova Science Publishers, Inc.
New York

Library of Congress Cataloging-in-Publication Data

Algae : ecology, economic uses and environmental impact / editors, Dagmar Krueger and Helga Meyer.
p. cm.
Includes bibliographical references and index.
ISBN 978-1-62081-280-8 (hardcover)
1. Algae--Ecology. 2. Algae--Economic aspects. 3. Algae--Environmental aspects. I. Krueger, Dagmar. II. Meyer, Helga.
QK568.E26A427 2012
561'.93--dc23

2012007264

Published by Nova Science Publishers, Inc. ✢ New York

CONTENTS

PREFACE

This book presents current research in the study of the ecology, economic uses and environmental impacts of algae. Topics include ultraviolet irradiation to control algal proliferation in the environment; alga Trachydiscus minutus as a new source of polyunsaturated fatty acids; systematics and taxonomic keys for the marine green algal family monostromataceae; the ecophysiology of soil algae; and an evaluation of the total phenolic content and antioxidant activities of crude extracts from red alga, Corallina elongata.

Chapter 1 - In the conservation field of buildings, historical monuments and subterranean environments, very few studies highlight the importance of finding alternative methods to chemical products in protecting against algal biofilm proliferation. Indeed, biofouling by microorganisms is a well-known biodeterioration phenomenon. Algae are considered to be significant contributors to the biodeterioration of stone, and that's why they are the subject of special interest in this chapter. Conservators of monuments or art works used many chemical applications to prevent or to control the algal proliferation on stone. But, due to the changes in European standards on their use, the chemical products must be replaced by environmentally friendly processes. Ultraviolet irradiation appeared to be a suitable process to control algal proliferation in these environments.

Chapter 1 Firstly describes problematic algal biofilms (proliferation and biodeterioration) on different substrata and also the effects of applied chemical treatments. Secondly, UV light and its effects on living organisms is defined and characterized. Finally, the experimental design and the obtained results on algae harvested in caves at the cellular and molecular scales (pigment contents, photosynthetic activity, DNA damages ...) and at the biofilm scale (proliferation measurements) is discussed. This study shows that UV treatment can be used as a conservation tool to eradicate algal biofilms, which proliferate in buildings and monuments.

Chapter 2 - A new bacterial isolate *Azospirillum brasilense* NH strain isolated from salt-affected soil in the north of Algeria was applied to inoculate durum wheat (*Triticum durum*, var. waha). This strain showed already an important positive impact on durum wheat (stem height and spike length) in absence of salinity. The strain shows also a slight effect under saline conditions. Moreover, the combination of this stain with the marine alga *Ulva lactuca* illustrated the best results in absence or presence of high saltiness (160 and 200 mM NaCl). Proline and total sugar production (stress parameters) under saline conditions is inhibited by using *A. brasilense* NH, but the both components are strongly inhibited when *U. lactuca* was added. Also, rates of germination, stem height per week, height/width of the largest leaf, roots and shoot dry weight, optical density (OD) of chlorophyll a and b and total rate in chlorophyll, stem and roots dry weight, 1000-seed weight, total spike length, spike dry weight, seed number per spike and total weight of seeds per spike, were positively affected. It was shown that this new strain restores entirely the vegetable growth at 160 and 200 mM NaCl. This study was carried out using a reference strain *A. brasilense* Sp7 in order to compare between the tow strains. It was shown that the reference strain Sp7 is totally inhibited at 200 mM NaCl.

Thus *A. brasilense* NH restored the vegetable growth of durum wheat and reduced the proline and total sugars under saline conditions in pot and field experiments. The results obtained proved the best effect of this new strain NH on durum wheat growth and inhibition of the osmotic stress. In additional to the restoration of durum wheat growth under saline conditions by *A. brasilense NH-U. lauca*, the algal extracts remains the good stimulator of plant growth in absence of stress.

The endophytic colonization of this strain was studied in presence or absence of salinity, in axenic conditions. Wheat roots were harvested and fixed with paraformaldehyde (4%) and hybridized using the FISH technique. Cofocal laser scanning microscopy (CSLM) was performed and the in situ localization of bacterial cells were obtained in the endophytic wheat roots. The best colonization was obtained at high salinity (200 mM NaCl). This result can justify the ability of this new strain to restore wheat growth at 200 mM NaCl. Thus, the results of the work displays that the combination of *A. brasilense* NH – *U. lactuca* extracts could be a good solution to solve the problem of salinity which is the serious threat hindering a wheat culture in arid and saline soils.

Chapter 3 - This chapter aims at testing the feasibility of developing an innovative micro-algae base technology that can concurrently produce biodiesel and protect the environment. The use of fossil fuel for transportation systems is costing the world billions of dollars each year and contributes to air pollution and the global warming. Microalgae have been recognized as one of the most promising source of biofuel. However, there is no local testing or commercial production of microalgal oil in many countries so far.

In this work, it is hypothesized that an economically viable microalgae-based biodiesel production system could be established if innovative photobioreactors were used along with suitable microalgal species rich in oil content. It is proposed to combine the biodiesel production with wastewater treatment and CO_2 reduction using microalgae. Such development will commence commercial feasibility projects. This technology is particularly useful in countries, where huge amounts of CO_2 are released to environment and huge demand on useful applications of wastewater.

In all countries, billions of dollars are spent each year for energy production. Many of the power plants consume diesel in the process of generating electricity and desalination of water. Most public transportation system uses diesel engines. It is well established that carbon dioxide emission from various transportation systems accounts for one third of the total carbon dioxide released into the atmosphere, leading to global warming. The toxic exhausted gas from vehicles also causes hazardous effects on human health.

Biodiesel, derived from vegetable oils or animal fats rather than petroleum, is an alternative fuel for diesel engines. It is renewable, non-toxic, and biodegradable. It can be used in existing diesel engines without modification, and can be blended in at any ratio with petroleum diesel. Currently, the major feedstock for biodiesel production is soybean oil. However, low oil production rates (e.g. 1 ~ 3 barrels of oil per acre of land per year) coupled with high production costs (e.g. feedstock accounts for 70-80% of biodiesel production costs) have limited the expansion of soybean-based biodiesel to meet the growing demand by the society. Therefore, it is important to develop more economically viable sources of feedstock for biodiesel production. Microalgae are known to exhibit 10- to 20-fold higher growth rates than agricultural crop plants, and certain microalgal species can accumulate large amounts of lipids or oil (30-60% of dry weight).

In this work is proposed in order to develop a complete prototyped model for the production and testing of biodiesel. The model will consist of cultivation reactor to grow suitable microalgae species, photobioreactors with wastewater and or CO_2 as feed to the algae, oil extraction from algae, chemical reactor to convert the extracted and finally test the produced biodiesel in a conventional diesel engine.

Chapter 4 - The alga *Trachydiscus minutus* (Bourr.)Ettl, strain Lukavský et Přibyl 2005/1, Eustigmatophyceae, is described here as a new source of polyunsaturated fatty acids (PUFAs), especially of eicosapentaenoic acid (EPA). These fatty acids (FA) are valuable food additives for man, animals and in aquaculture.

T. minutus was first described as *Pseudostaurastrum minutum* by Bourrelly (1951), from a small fishpond in France, in 1951. It is now found in various water localities, e.g. ponds, lakes and streams, but usually as a rare species. Its principal mode of reproduction is by autosporogenesis, and in prolonged darkness, zoospores are produced. Only chlorophyll_*a* was evaluated; other pigments are violaxanthin, vaucheriaxanthin ester, and ß–carotene.

In laboratory cultures, biomass can reach 16 $g.l^{-1}$ dry weight (DW), consisting of 26-31% total lipids (TL) and EPA comprises up to 42% of TL. The TL content can be increased to 39% of DW by nutrient starvation (N,P,S deficiencies) resulting in an EPA content of up to 35.8% DW, and a productivity of 88 mg $l^{-1}.d^{-1}$. Targeted stereoisomers of triacyglycerols (TAGs) can also be obtained. Oil drops are prominent and an iodine value of about 216 was determined by Raman microspectroscopy, even in individual cells; this has high potential value for the food industry. *T. minutus* also contains 26% of saturated myristic fatty acid, and about 48% protein suitable for food and fuel. The optimal growth temperature was about 27 ^{o}C, but it survived below 15 and at 35 ^{o}C. Extreme temperatures resulted in increased TL content, up to 35% DW, and decreased carbohydrate (13.6% DW) and protein (30-38% DW). High temperature elevated the activity of superoxide dismutase, catalase and proteases, but glutathione reductase, and peroxidase activity decreased. Under 15 ^{o}C, all enzymes except glutathione reductase were inhibited. In the laboratory, optimal growth occurred under continuous light of 2x132 μmol $photons.m^{-2}.s^{-1}$, whereas the maximum light levels evaluated in the greenhouse were up to 640 μmol $photons.m^{-2}.s^{-1}$.

In an orbicular cultivation unit inside a greenhouse in N.Greece (vol. 200 litre, area 2 m^2) *T. minutus* produced 0.6 $mg.m^{-2}$/day i.e. 3.4 $mg.l^{-1}.day^{-1}$ of EPA. In SW. Bohemia, using a thin-layer inclined photobioreactor (volume 150 litre, area 15m^2) in a greenhouse, *T. minutus* produced up to 0.25-0.45 $mg.m^{-2}.day^{-1}$, i.e. 20-30 $mg.l^{-1}.day^{-1}$ of EPA. *T. minutus* proved to be tolerant of manipulation, e.g. centrifugation, oscillations of light, temperature and CO_2 concentration, contamination with another algae etc. Sedimentation followed by centrifugation and lyophylization, or drying in the sun on plastic foil was suitable for harvesting. Flocculation also has potential (Přibyl pers. com.) and no parasites have yet been observed in *T. minutus.*

T. minutus was successfully immobilized in agar or alginate. Its potential for further research and exploitation is high and comparable with, or better than other algae used for biotechnological production of PUFAs, and especially EPA.

Chapter 5 - Marine green algal family Monostromataceae consists of single cell-layered green seaweeds distributed throughout the world, yet identification of it being one of the most challenging. Monostromatic green seaweeds are all edible and are commercially cultivated in Japan for centuries. A thorough understanding of its systematics is indeed a necessity for any empirical investigation. Working dichotomous taxonomic key for this family is presented for the first time for the ready benefit of field phycologists. Recent advances in the systematics of this family with a special emphasize on the Japanese varieties are also presented.

Chapter 6 - In most biological textbooks the term "alga“ is used to designate a group of photosynthetically active organisms that are bound to aquatic habitats and lack the classical differentiation into root, stem and leaves. Only experts are aware of the fact that algae are very successful competitors in terrestrial and aerial habitats as well and that they can survive also under aphotic conditions. Accordingly, there is ample information available on taxonomy, physiology and ecology of aquatic algae and on their application in biotechnology whereas on aeroterrestrial algae comparably few data exist. This is the more astonishing since aeroterrestrial algae have to cope with much more diverse and quickly changing environmental stressors than aquatic algae do that live in a more or less constant environment. Aeroterrestrial algae are specialists for interfaces, they settle e. g. substrates exposed to the air such as soil, tree barks, rocks, fences, roofs, facades and rubber seals of car windows. There environmental conditions such as light, UV, temperature and moisture can change quickly. What is more, those algae contribute a significant part to CO_2-fixation in their ecosystems, the exact amount of which still needs thorough investigation. Thus the study of the ecophysiology of aeroterrestrial algae, of their adaptive features and their potential biotechnological applicability is one of the most promising fields in current phycological research.

In this chapter an overview will be given on what is known on the ecophysiology of soil algae, i. e. on that part of the aeroterrestrial algae that lives in soil, and on current research on their environmental impact as well as on their potential use in biotechnology.

Chapter 7 – This chapter evaluates the total phenolic contents and antioxidant activities of crude extracts from a red alga, *Corallinaelongata*. The phenolic compounds in the extracts have been identified and quantified by reverse-phase high-performance liquid chromatography. Phloroglucinol was by far the predominant polyphenol found in the extracts, followed by catechin and epicatechin. Extracting solvents such as water, methanol and water/methanol (1/1) significantly affected the total phenolic amounts determined using the Folin-Ciocalteu method. The total phenolic contents ranged from 443.0 to 487.4 mg gallic acid equivalents per 100 g of dry plant material. The antioxidant efficiency of the extracts was evaluated by measuring the ability to scavenge the stable free radical 1,1-diphenyl-2-picrylhydrazyl, finding that it ranged from 37.0 % to 66.4% (inhibition percentage). The high correlation between the yield of extraction, the total phenolic content and the radical scavenging activity was found, allowing us to conclude that phenolic constituents are mainly responsible for the observed antioxidant activity in the extracts. Because of the wide range of biological activities of phloroglucinol and the other polyphenols detected in the extracts, *Corallina elongate* appears to have many potential industrial uses.

In: Algae
Editors: Dagmar Krueger and Helga Meyer
ISBN: 978-1-62081-280-8

Chapter 1

UV IRRADIATION AS AN ALTERNATIVE TO CHEMICAL TREATMENTS: A NEW APPROACH AGAINST ALGAL BIOFILMS PROLIFERATION CONTAMINATING BUILDING FACADES, HISTORICAL MONUMENTS AND TOURISTIC SUBTERRANEAN ENVIRONMENTS

Fabien Borderie[1], Laurence Alaoui-Sehmer[1], Faisl Bousta[2], Geneviève Orial[2], Dominique Rieffel[1], Hervé Richard[1], and Badr Alaoui-Sosse[1*]

[1] Laboratoire Chrono-Environnement (LCE), Université de Franche-Comté, France
[2] Laboratoire de Recherche des Monuments Historiques (LRMH), France

ABSTRACT

In the conservation field of buildings, historical monuments and subterranean environments, very few studies highlight the importance of finding alternative methods to chemical products in protecting against algal biofilm proliferation. Indeed, biofouling by microorganisms is a well-known biodeterioration phenomenon. Algae are considered to be significant contributors to the biodeterioration of stone, and that's why they are the subject of special interest in this chapter. Conservators of monuments or art works used many chemical applications to prevent or to control the algal proliferation on stone. But, due to the changes in European standards on their use, the chemical products must be replaced by environmentally friendly processes. Ultraviolet irradiation appeared to be a suitable process to control algal proliferation in these environments.

Firstly in this chapter, we describe problematic algal biofilms (proliferation and biodeterioration) on different substrata and also the effects of applied chemical treatments. Secondly, we define and characterize UV light and its effects on living organisms. Finally, we present the experimental design and the obtained results on algae harvested in caves at the cellular and molecular scales (pigment contents, photosynthetic

* Corresponding author, Tel. +33 (0)381 665 730, Fax. +33 (0)381 665 797, E-mail: *badr.alaoui-sosse@univ-fcomte.fr*.

activity, DNA damages ...) and at the biofilm scale (proliferation measurements). This study shows that UV treatment can be used as a conservation tool to eradicate algal biofilms, which proliferate in buildings and monuments.

1. INTRODUCTION

Many constructions (houses, buildings ...), historical monuments (churches, crypts, abbeys ...) and subterranean natural cavities (caves, chasms ...) are subject to biofouling by algae for different reasons. Whether due to environmental (exposure, humidity, nutrient contents ...), anthropic activities (artificial lighting, CO_2 emissions by breathing) or related to the intrinsic characteristics of the support (porosity, mineral composition ...), algae proliferate very rapidly in the form of biofilms, and can affect the longevity of the substrate. Moreover, they very often constitute an aesthetic nuisance when they grow on art works or on monuments. For several decades, conservators have used chemical products in order to control algal proliferation. These products appear to be very efficient but the main disadvantage of these compounds is their toxicity against most organisms and their persistence over the years. The result is a pollution phenomenon caused by these toxic substances that can be transported around the target area. Especially in caves or chasms, these products are lixiviated and the whole of subterranean ecosystem can be contaminated. Following the awareness of the danger of using these kinds of substances and also for the sake of complying with new European standards, chemical treatments have to be replaced by environmentally friendly processes.

Since 2007, the *Laboratoire de Recherche des Monuments Historiques* (LRMH) in collaboration with the *Laboratoire Chrono-Environnement* (LCE) have worked together on a new process that consists of the use of UV irradiation as a physical and alternative method to chemical treatments.

For a few decades, because a stratospheric ozone layer depletion was detected, several authors have studied the effects of ultraviolet radiations, part of the solar light, on living organisms in laboratory or in open-air conditions mainly in marine habitats (Begum et al. 2009; Hsu et al. 1998; Jayakumar et al. 1999; Karsten 2008; Ma and Gao 2010; Menetrez et al. 2010; Moharikar et al. 2006; Moody et al. 1999; Sinha and Häder 2002). This ozone layer constitutes the only efficient protection against ultraviolet radiation (mainly UV-B and UV-C) generated by the sun. These harmful radiations may cause significant damage to plants, animals or human populations. These authors were shown that UV-radiation causes deleterious effects on vital metabolic functions such as photosynthetic activity, cellular division and DNA replication or germination spores and hyphal proliferation. Moreover, UV-radiation can also affect proteins or cell membranes and also the integrity of the nucleus apparatus, causing the induction of Programmed Cell Death (PCD).

Nowadays, UV treatment is largely used in several domains. Firstly, UV irradiation is used in semi-closed environments such as hospitals to disinfect surfaces or chirurgical tools. Indeed, this treatment aims to eradicate microorganisms, which can be responsible for infectious diseases (bacteria and fungi). (Menetrez et al. 2010) have shown that 180 seconds of UV-B cause 81% of mortality on *Bacillus* spores. Secondly, UV irradiation is also used in water treatment to eliminate bacteria or fungi spores. For example, (Shirayama et al. 2001) has shown that hydrocarbons are degraded under UV irradiation ($23J.min^{-1}.cm^{2}$) before conventional treatment in the wastewater treatment plant. Finally, this treatment is used in the food-industry to increase the time of storage without contamination of ready-to-eat meat products. Nowadays, a broad range of applications exists, but this process has never been used or studied for application in the material or monuments conservation field. However, (Grobbelaar 2000) has suggested the use of UV-C lamps against lithophytic algae, which grow on the walls of Cango Caves (South Africa), when chemical treatments would not be efficient.

In this chapter, we will firstly describe the biofouling phenomenon by algae. Where does it occur? What are the types of species present? What are the problems with the biofouling phenomenon? Then, we will discuss the chemical products used to fight against algal biofilms and about their deleterious effects. Secondly, we will describe the UV light characteristics and their effects on living organisms. We will present some conducted experiments and obtained results on the effects of UV-C radiation on algae harvested in caves (effects on physiological and molecular parameters, effects at the biofilm scale). Finally, we give an example of the efficacy of the treatment applied to the wall of a historical monument.

2. Biofouling by Algae

All constructions, monuments and art works can be contaminated by living organisms (algae, cyanobacteria, fungi, mosses, lichens and higher plants). Following the ecological successions, unicellular and multicellular organisms like algae and cyanobacteria are generally pioneers in the biofouling phenomenon. On the nutrient-depleted stone surfaces they build up a visible protective biofilm enriched with inorganic and organic biomass allowing the set up of other microorganisms and higher plants (Nugari et al. 2009). Because they are the first link of the biodeterioration of monuments and art works, only the biofouling caused by algae will be studied in this chapter.

Biological disturbances by algal biofilms occur all around the world. Churches of South America cities, places in South Asia, building facades in a great number of world cities and caves in Europe, Australia or the African continent are subject to biofouling by algae. Given the broad range of structures where algae proliferate, the conditions conducive to their growth are multiple. Briefly, humidity rate, illumination, nutrient contents and characteristics of substrate (Barberousse et al. 2007; De Muynck et al. 2009) are the predominant factors influencing the algal development. The term of "bioreceptivity" defines the aptitude of a material to be colonized by living organisms (De Muynck et al. 2009; Guillitte 1995) and in the case we are interested in, the required support properties that favor the algae installation. In fact, the colonization of a substrate by algae is related to some key factors such as surface roughness, moisture content, chemical composition, structure and texture of the material (De Muynck et al. 2009; Guillitte 1995). Bioreceptivity is a well-known phenomenon encountered both on building facades, on historical monuments and in subterranean environments such as caves. Nutrient contents are low in subterranean environments but sufficient for algae growth as long as artificial illumination and humidity are maintained (Grobbelaar 2000). On historical monuments, these conditions are similarly conducive on building facades.

Algal cells or filaments have the capability to merge to form a unit called biofilm. The extracellular polymeric substances (EPS) secreted by cells in a biofilm allow adhesion to the surface and facilitates the entrapment of airborne particles and organic compounds. This entrapment of particles causes a dirty appearance of the substrate (Scheerer et al. 2009) and may cause the loss of touristic attraction by masking the interesting parts of monuments or art works. Biofilms also cause possible physico-chemical deterioration of the substrate (see paragraph 2.2).

In the following paragraphs we will present some examples of biofouling in different environments and implying different algal species.

2.1. Species, Concerned Areas and Factors of Colonization

As said previously, the areas concerned by the biofouling problem include building facades and historical monuments, and also with a particular attention, the subterranean environments. In the examples presented below, the characteristics of the support (nature of stone, architectural characteristics, biotic and abiotic factors ...) can mostly explain the algal colonization. The species found in biofilms are extremely diverse with a predominance of Chlorophyta. Among the Chlorophyta, several genus may participate in biofilm formation. The species mentioned below therefore do not constitute an exhaustive list of the algae involved in the biofouling.

2.1.1. Buildings Facades

The biofouling by algae on building facades is a widespread phenomenon which affects constructions all around the world. However, the situations to which we will refer in this paragraph occur especially in Europe and in South America. The facades of many constructions could constitute conditions conducive to facilitate the algal development. Indeed, the presence of architectural defaults like exposed frontages, leaking gutters, air-conditioning vents or water rejection on bases of walls (Barberousse et al. 2006) fulfil the conditions necessary to the development of algae. (Barberousse et al. 2006) showed that it was a ubiquitous problem in the quasi totality of France regions. They identified a broad range of algal species implied in the biofouling process. The most common alga found on building facades in France is *Klebsormidium flaccidum*. *Klebsormidium* is a cosmopolitan genus of filamentous green algae, widespread in terrestrial and freshwater habitats and dominant in biofilms (> 50%). This alga has been reported in rivers, bogs, on tree bark, stone monuments and bases of walls (Rindi et al. 2008). Its capacity to survive in a highly dehydrated state (Barberousse et al. 2006) is at the origin of its large occurrence even if it generally grows in places with high humidity. The second most common genus after *Klebsormidium* is *Trebouxia,* which may represent as many as 4% of the algal cells forming the biofilm. According to the study of (Barberousse et al. 2006), other species are also identified on building facades: *Apatococcus lobatus*, *Bracteacoccus* sp., *Chlorella ellipsoidea*, *Chlorella minutissima*, *Desmococcus olivaceus*, *Keratococcus bicaudatus* ...

(Gaylarde and Gaylarde 2000) have studied and identified the composition of phototrophic microorganisms in biofilms on painted buildings in Latin America towns (Argentina, Bolivia, Brazil, Mexico and Peru). More than 46 genera of algae - and more for the cyanobacteria - were identified and the algal genus *Chlorella* showed the most widespread occurrence (> 68%). On this surface, organisms are specialized because they are exposed to extreme conditions generated by the repeated desiccation due to the low water retention of paints as well as intense UV irradiation, especially at high altitude in the tropics, and wide swings of temperature. (Barberousse et al. 2006) studied the factors involved in the colonization of building facades by algae and cyanobacteria. Temperature is the main environmental parameter followed by humidity content, exposition and precipitation (Barberousse et al. 2006). The building façade characteristics also play an important role in the colonization by microorganisms: the surface irregularities, porosity and roughness facilitate the retention of water, which is a necessary step for the installation of microorganisms.

On the other hand, the installation of algae in such extreme environment is also due to some special physiological adaptations. (Karsten et al. 2007) compared some physiological parameters of green aeroterrestrial algae, *Stichococcus* sp. and *Chlorella luteoviridis*, both common in biofilms from facades, to two green algae growing on soil, *Myrmecia incisa*, or in brackish water, *Desmodesmus subspicatus*. The results show that the success of biofilms of algae is at least, for a great part, due to their ability to synthesize protective sunscreen compounds, the so-called mycosporines-like amino acids (MAAs). In fact, the building facades are exposed to high solar radiation (UV-A and UV-B, see paragraph 3), and it was shown that the aeroterrestrial green algae were able to synthesize MAAs whereas green algae from soil or brackish water were not. Thus, the presence of MAAs involves a high UV-tolerance thus preserving growth and photosynthetic performance, which allow aero terrestrial green algae to survive in these environmentally harsh habitats.

2.1.2. Historical Monuments

Like buildings, historical monuments are also subject to biofouling by algae. The conditions conducive to the installation of algal biofilms are the same as those on building facades *i.e.* humidity content, structure and texture of support (roughness and porosity), exposition and defaults in the support that may cause micro-humidity areas. A good example of that biofouling by algae on historical monuments is the case of the Lions Fountain at the Alhambra Palace, Granada (Spain). The biodeterioration of this fountain was studied by (Sarró et al. 2006). Indeed, part of this fountain is fouled by an algal biofilm essentially composed of Cladophora, Chaetophorales, Chlorella and Chlorococcum genera, all genera and species also commonly described on facades. Cladophora is a green filamentous alga with rhizoids that can penetrate into the stone and deteriorate it. This alga presents a gelatinous layer that swells in the presence of water and can cause damage of the support by fracturing and disintegrating the stone. In the case of fountains, the permanent humidity and splashing as well as an exposition to climatic factors (rain, sun ...) contribute to facilitate the formations of biofilms with photosynthetic microorganisms such as algae and cyanobacteria.

(Tomaselli et al. 2000) have analyzed the biodiversity of photosynthetic microorganisms growing on stone monuments such as fortresses, basilicas, churches and statues in several Italian cities. They used morphological characterization and molecular techniques to identify the genera and species. In their investigations, the authors have found two dominant taxa of photosynthetic microorganisms: cyanobacteria and chlorophyta. Among the chlorophyta, *Ulothrix*, *Chlorella* and *Stichococcus* were the most widespread genera. In comparison between their study and literature, the authors have tried to highlight relationships between the organisms and the nature of the substrate but no clear relationship could be established. They have nevertheless shown that the most common chlorophyta, *Klebsormidium flaccidum*, is a cosmopolitan widespread species that has been found on all lithotypes (artificial, calcareous and siliceous). On the contrary, *Stichococcus* and *Apatococcus* occurred only on calcareous substrates. Apart from these two cases, there is no correlation between the algae species presence and type of substratum.

2.1.3. Subterranean Environments: Example of Caves

Subterranean environments such as caves are particular ecosystems with abiotic and biotic factors that define a fragile dynamic balance (low temperatures, high humidity and carbon dioxide contents. After their discovery, a significant number of caves are modified to permit tourist visits. Indeed, walkways and illumination systems are established to show karstic formations (stalagmites and stalactites) and, possibly, prehistoric art works like mural paintings and rock carvings. The most famous example of a show cave is the Lascaux's cave (Dordogne, France). Since its discovery in 1940, Lascaux's cave has received thousands of visitors per year. In 1962, a major biological disturbance led to the closure of the cave to the public. The presence of major algal biofilms on mural paintings, named "the green sickness" or "lampenflora" (Lefèvre 1974) threat the integrity of prehistoric paintings and also the karstic formations. Lefèvre and Laporte (1969) have identified the green algae *Bracteacoccus* sp. (unicellular green algae, Chlorophyta) as the principal agent of this contamination. The reason of this brutal proliferation is the over-exploitation of the cave, which leads to a panel of consequences:

(i) the CO_2 content of air was increased by the visitors breathing and was then utilized by the algae to improve their photosynthetic activity,
(ii) the lighting system used to showcase the paintings provides a light source essential to establishment of the photochemical phase of photosynthesis,
(iii) and furthermore, visitors may have been a vector to algal spores from outside.

As mentioned above, visitors and illumination systems are predominant factors to the installation of algae in a cave. But others important factors have to be taken into account, such as the humidity content being close to 100%, the carbon dioxide content and the charged-HCO_3 water, which led to sufficient nutrients.

(Grobbelaar 2000) has analyzed several species or genera of green algae, which proliferate in the Cango Caves, one of the most important tourist sites in South Africa. It appears that the wettest areas were dominated by *Chlorella ellipsoidea* and maybe by some *Apatococcus* (filamentous green alga) and that on the contrary, *Phormidium* and *Symploca* genera have been identified in the driest areas. In these closed environments, algae can be entirely desiccated and if conditions become favorable again, they are able to start growing again. Furthermore, like Chlorella, the algae of subterranean environments are highly shade acclimated and possess a very low compensation point for light. Furthermore, according to (Grobbelaar 2000), algae present in the Cango Caves are in a good physiological state, precisely because they are able to take advantage of all the photons even at very low light intensities.

In the Crypt of Original Sin (Matera, Italy), (Nugari et al. 2009) have analyzed the biodeterioration phenomenon on mural paintings. The Crypt of Original Sin is a rocky habitat with Byzantine style paintings, belonging to the wide cultural heritage of Matera. The configuration of the crypt conducts a microclimate and an ecological distribution of microorganisms, similar to semi-enclosed environments. Indeed, high densities of algae (*Apatococcus lobatus*, *Cholrella vulgaris* and *Chlorococcum sp.*) have been detected in the most humid areas.

Some species of the subterranean environment are the same as those found on buildings and on historical monuments but the physiological characteristics that allow their establishment on the monuments are not the same as those required for underground environments. A successful implantation on historical monuments and building facades implies that algae are able to synthesize sunscreen compounds like MAAs (Karsten 2008; Rozema et al. 2002; Ryan 1992; Sinha et al. 2007) conferring them an UV-protection. On the contrary, algae in caves have to be shade-acclimated. And most of them are so well adapted that an excess of light may induce photoinhibition (Altamirano et al. 2004; Hanelt et al. 2006), which causes a decreased photosynthetic activity and an increased cellular mortality.

2.2. Biodeterioration Phenomena

Biodeterioration phenomena caused by photosynthetic microorganisms on stone are multiple. In this paragraph, we briefly describe the principal processes encountered in the major cases (physical and chemical processes). Because the chemical reactions at the interface between microorganisms and stone are very complex, and because it is not the aim of this work to detail that kind of chemical reaction, we will limit ourselves to the description and to comment on a few examples.

Biofilms of algae may cause the dirty appearance of the substratum both by the presence of EPS on the surface of biofilms (Scheerer et al. 2009) and by the presence of organic colored pigments in microorganisms (the photosynthetic pigments, chlorophyll a and b, are for example, responsible of the green color). In addition to the problem of a dirty appearance, the substratum might be deteriorated. Under the term biodeterioration are regrouped two basic mechanisms: the biophysical mechanisms, which are determined by the porosity or shape of the substratum, and the biogeochemical mechanisms which in turn are controlled and determined by chemistry of minerals (Scheerer et al. 2009; Warscheid and Braams 2000).

Algae cause mainly biophysical damages on the substrate. Indeed, as described previously in the case of the Lion Fountains, filamentous algae, such as Klebsormidium, can penetrate within the fissures of rocks and may cause biophysical damages. The water uptake by the filaments and the expansion of the cell mass exert a pressure within the fissures that led to the detachment or spalling of the superficial layer of the structure of the substrate. These wetting and drying cycles and subsequent expansion and contraction of the cell mass, lead to changes on the pore size, dry density, water content, surface hardness and act as a permeability block for the evaporation of humidity within the stone. Moreover, algae also cause biogeochemical damage like carbonatation, a consequence of the removal of CO_2 during photosynthesis and the precipitation of carbonate. Thus, the carbonated crust affects the structure of the substrate (Bolivar and Sanchez-Castillo 1997). Furthermore, the excretion of corrosive organic acids especially on limestone and marble and the uptake and accumulation of sulphur and calcium into their cells, lead to the loosening of stone particles.

2.3. Chemical Treatments against Biofouling

There are many solutions to fight against the problem of biofouling, depending on the types and the size of areas that have to be treated. Moreover, the treatments are not the same on historical monuments, on art works or in subterranean environments. Historical monuments and art works are periodically subjected to operations of restoration when the stone starts to present too many degraded areas. The restoration operations consist of the reconstruction of parts of stone. In this phase of reconstruction, conservators include in the material chemical products to prevent the biofouling by microorganisms. But, this kind of restoration operation is not possible in the subterranean environments and the only treatment options are curative treatments. In this case, chemical products, liquid or powder (biocides, algaecides, fungicides ...), are directly applied or pulverized on the stone to eradicate microorganisms.

The preventive strategy is a well-controlled method. The risk of pollution is reduced because chemical products are included into the stone (cement, mortar, concrete ...) and are not leached around the target area. In a recent study (Alum et al. 2008), the authors have demonstrated that the incorporation of zinc oxide and ammonium bromide into the cement is effective in preventing biofouling by algae in water distribution canals. A neutral pH value of 7 associated with high summer temperatures promote the stability of zinc oxide and minimizes its water solubility and, therefore also the pollution risk. Another preventive technique is to treat the surface of the material with so called water-repellent products. This technique is currently used in the building constructions. The major role of a water-repellent is to prevent liquid water from penetrating into the stone, by capillary action, and thus to limit biological contamination by moisture scarcity (Moreau et al., 2008).

The curative strategy is applied when the stone surfaces are already contaminated by microorganisms. A large broad of chemical products are used in this case, for all kind of microorganisms (algae, cyanobacteria, fungi, lichens and mosses).

While the first strategy is well controlled with a limited risk of pollution, pulverization of biocides on stone is a less environmentally friendly process, because chemical products can be leached around the target area and cause pollution phenomena, especially in the sub-terranean environments where groundwater can be polluted by infiltration in the karstic web. For example, in the Lascaux's caves during the campaign of chemical treatments against fungi contamination, quaternary ammonium salts (fungicide and bactericide) were trapped into the clay layers and thus degrade much more slowly. This product causes the lysis of the cellular membrane and the denaturation of proteins and enzymes. Thus, the impact of this product on the environment can be irreversible. (Tretiach et al. 2007) have tested the efficacy of a biocide (Koretrel©) on photoautotrophic organisms contaminating rock samples in open-air conditions. The results show that all organisms present before treatment showed evident signs of damage, such as changes in color and cell death after treatment with biocide. Treatments are thus efficient but the problem of persistence should not been neglected in such fragile environments.

(Nugari et al. 2009) and (Grobbelaar 2000) have compared the effects of several biocides on microorganisms such as algae, fungi and cyanobacteria. The first study concerned the biological colonization of Byzantine mural paintings in rocky habitats (Crypt of original Sin, Matera, Italy). The authors have selected squares of painted-on stone surfaces and have applied different biocide solutions (Rocima 110©, Preventol R80© and Umonium 38©).

These products are the most commonly used in the surface treatment against biofouling. Rocima and Preventol appear to be more efficient than Umonium against all types of biofilms.

In the second study, (Grobbelaar 2000) has studied the effects of two herbicides (Atrazine and Simazine) and of hydrogen peroxide [H_2O_2]) on lithophytic algae growing in the Cango Caves (Swartberg Mountains, Western cap region, South Africa). These three products completely inhibited algal growth in the laboratory tests, but H_2O_2 is preferentially adapted to the caves because it's totally environmentally friendly and dissociated into water. On the contrary, Atrazine and Simazine are not suitable in caves because ethanol and acetone fumes made the atmosphere unpleasant in the caves. Moreover, these compounds are forbidden in France and European Union since 2000 due to their environmental impact and particularly their persistence over the years in the ecosystems.

Chemical products appear to be very efficient in the case of curative treatment, but alternative methods should be developed to limit their use in environments with high tourism value and to preserve the ecological balance of ecosystems.

The UV irradiation as a physical and germicidal treatment could be a suitable alternative method of treatment to control or limit biofouling. This technique is already used in several domains such as medical, food industry or water treatment to disinfect and eliminate microorganisms. The aim of the following paragraphs is to describe the ultraviolet light and its effects on living organisms.

3. The Ultraviolet Light

3.1. Definition and Characteristics

Ultraviolet light (UV) is an electromagnetic radiation mostly classified as non-ionizing except for short wavelengths, around or shorter than 150 nm. The UV light is a part of the solar spectrum with wavelengths comprised of visible light and X-rays. The UV radiations are divided into three categories depending on their wavelengths: UV-A (315-400nm), UV-B (280-315nm) and UV-C (100-280nm). Unlike UV-A and UV-B, the UV-C are the only rays entirely filtered by the ozone layer. This is fortunate because they also represent, with their highly energetic photons, the part of the solar spectrum that is the most harmful for living organisms.

3.2. Effects on Living Organisms

Since the middle of the seventies, consecutive to the detection of the ozone layer depletion, many studies have analyzed the effects of ultraviolet radiation on living organisms (Feng et al. 2007; Hodoki 2005; Xiong 2001; Zeeshan and Prasad 2009). In fact, the ozone layer is the part of the earth's atmosphere that plays the major role in the filtration of solar radiation. Consecutively, several authors have studied the effects of UV radiation on micro and macro-algae, dinoflagellates, cyanobacteria, bacteria, fungi and higher plants in laboratory conditions as well as in open-air conditions (Begum et al. 2009; Hsu et al. 1998;

Jayakumar et al. 1999; Karsten 2008; Ma and Gao 2010; Menetrez et al. 2010; Moharikar et al. 2006; Moody et al. 1999; Sinha and Häder 2002).

The different effects reported *i.e.* damages or responses caused by ultraviolet radiation on living organisms are summarized in Table 1.

According to (Jenkins 2009), the effects of UV irradiation may vary greatly depending on the used wavelength, the quantity of received energy and the duration of the experiment. On one hand non-specific effects are observed including production of ROS, DNA damage and the synthesis of jasmonate salicylic acid and ethylene. On the other hand more specific effects may appear as a consequence of a UV photoreceptor stimulation whose existence has yet to be proved. The latter would be responsible for the modified morphogenesis and histology currently observed on plants under UV exposure. Anyway, almost all cell components may constitute a target in case of UV exposure. The following discussion will summarize the most common.

Firstly, histological changes have been observed on some algae strains or higher plants. Working on a macrophyte, *Juncus effusus*, (Najeeb et al. 2011) showed that the mesophyll cells presented several morphological changes after UV-C treatment, including a reduced mesophyll cell size, an increased cell wall thickness, a disruption of the thylakoid membrane and an unusual accumulation of plastoglobuli in chloroplasts. (Moharikar et al. 2006) have shown that the unicellular green algae *Chlamydomonas reinhardtii* shows morphological changes under various doses of UV-C. The control cells showed a normal centrally placed nucleus with condensed chromatin uniformly distributed and dense cytoplasm. In contrast, the cells exposed to UV-C showed high vacuolation and extensive chromatin fragmentation.

Beside these histological modifications, the physiology of the cell is also greatly affected under UV irradiation. A large part of the observed symptoms may be linked directly or indirectly to ROS (Reactive Oxygen Species) production. It has been shown that the ultraviolet radiations, at least those whose wavelengths are the shortest, are at the origin of an oxidative stress. The high energy photons in fact, generate the formation of Reactive Oxygen Species (ROS) such as superoxide anion ($O_2^{\cdot-}$), hydrogen peroxide (H_2O_2) and single-state oxygen ($^1O_2$) (Ma and Gao 2010). ROS, due to their high reactivity, are known to attack and to degrade almost all biomolecules and particularly lipids from biological membranes, proteins and DNA (Ma and Gao 2010) and may damage the photosystem II reaction center leading to the decrease of photosynthetic activity and viability loss (Janknegt et al. 2007). However, living organisms have developed an antioxidant system to minimize the deleterious effects of ROS. This system consists of several enzymes scavenging ROS (superoxide dismutase, ascorbate peroxidase, glutathione peroxidase and other peroxidases, catalases…), in enzymes able to detoxify lipid peroxidation products (e.g. phospholipid-hydroperoxide glutathione peroxidase) and in a network of low molecular mass antioxidants (e.g. ascorbate, glutathione, phenolic compounds and tocopherols). It assists in controlling the level of ROS and protecting cells against oxidative stress (Blokhina et al. 2003; Mahdavian et al. 2008; Shao et al. 2008).

Some of the major indirect effects of the ROS formation under UV exposure are the damage caused to DNA, to proteins, to the membrane lipids and the inducing of Programmed Cell Death (PCD).

The damages caused on nucleic acids, especially on DNA consist mainly in the formation of Cyclobutane Pyrimidine Dimers (CPDs), considered as toxic and mutagenic photoproducts

(Moharikar et al. 2006). In that way, UV-induced damage can significantly compromise the accuracy of nucleic acid transcription and replication, causing misreading or erroneous replication, leading to an increasing number of mutations (Karsten 2008).

Damage caused to protein mainly arises because of the UV absorbing amino-acid present in their structure. But it can also lead to the rupture of disulfide bonds within the proteins itself (Bischof et al. 2007).

Concerning lipids, UV exposure leads to lipoperoxidation, inducing formation of peroxy radicals than maintain the reaction (Bischof et al. 2007).

Finally, UV-C irradiation is also known to induce Programmed Cell Death (PCD) or apoptosis at high as well as at low doses. Apoptosis is a genetically controlled process (e.g. during the growth) or may be induced by several cell signals under stress conditions such as depolarization of the mitochondrial membrane, chromatin condensation, DNA fragmentation, induction of caspases and ROS accumulation…(Danon and Gallois 1998; Grütter 2000; Moharikar et al. 2006; Vardi et al. 1999). (Moharikar et al. 2006) have analyzed the apoptotic-like cell death pathway in *Chlamydomonas reinhardtii*, a unicellular green alga, induced by UV-C irradiation. Many morphological changes were observed (e.g. highly condensed chromatin and increased vacuolation) accompanied by damages on nuclear apparatus (e.g. nucleosomal laddering and TUNEL positive cells) and induction of caspases-3-like epitopes, a precursor in PCD.

Concomitantly, in photosynthetic organisms, the photosynthesis itself constitutes a prime target for UV radiations. Because this metabolism is the main cellular metabolism, any modification of its functioning affects the overall cell metabolism.

In case of photosynthetic organisms, from Cyanobacteria to algae and higher plants, the UV radiation has an impact on photosynthetic activity, pigment contents and cellular viability. Among the chloroplast's membrane complex and the photosystem's proteins, the protein D1, which constitutes the core of the photosystem II (PSII) from photosynthetic organisms, is especially sensitive to UV. Its degradation rapidly leads to a reduction of PSII efficiency, i.e. of the photosynthetic capacity. In the same way, the key enzyme of the carbon fixation metabolism, RubisCO is affected (Karsten 2008). Moreover, still regarding these photosynthetic organisms, ROS cause damages to chlorophylls and phycobilins, the pigments of photochemical electron transport systems. Under high doses of UV-A or UV-B, photosynthetic pigments such as chlorophyll a and b are bleached and their quantities are decreased with consequences on photosynthetic activity (Yakovleva and Titlyanov 2001). Thus, photosynthetic activity is affected and the viability of cells is severely compromised.

Moreover, several authors have studied the effect of UV radiations on photosynthetic performances by measuring the chlorophyll fluorescence. Chlorophyll fluorescence is a dissipative way for unused excess energy to drive photosynthesis. Hence, the measure of chlorophyll fluorescence gives an indication about photosynthetic performances and particularly the efficiency of PSII (optimal quantum yield of PSII or Fv/Fm) and the electron transport rate (ETR) (Maxwell and Johnson 2000). (Gómez et al. 1998) have studied the effect of different wavelengths of solar radiation including UV-A and UV-B on the green alga *Dasycladus vermicularis* under outdoor conditions. Their results show that the Fv/Fm is reduced by 30% and 54%, respectively, under UV-A and UV-B compared to the control values. The authors have also analyzed the total recovery of Fv/Fm values after treatments. They showed that under UV-A the complete recovery is reached after 2h while under UV-B, complete recovery is reached after 24h. According to them, these results can be explained by two assumptions: the more pronounced effects of UV-B radiations compared to UV-A, and the possible synthesis of sunscreen compounds, perhaps MAAs, which are absorbed mainly within the UV-A region and which are less efficient to counteract UV-B radiations. To cope with these deleterious damages on vital metabolic functions, some photosynthetic microorganisms, algae and cyanobacteria, in aeroterrestrial, freshwater or especially marine environments, have developed various strategies. These include avoidance, numerous protective mechanisms and repair of essential biomolecules (Karsten 2008). A physical move away from harmful UV-B is one of the most effective avoidance strategies. In marine environments, algae and cyanobacteria thus migrate vertically to low radiation habitat like sediments. Another similar example is the case of endolithic cyanobacteria, which are known to live inside mineral crusts or rocks to avoid high radiation. Most photosynthetic organisms, such as algae and cyanobacteria, have evolved a complex defense system against ROS, including non-enzymatic antioxidants like carotenoids (Karsten 2008). And it is well known, as it was underlined by (Stahl and Sies 2003) that the role of carotenoids in algae and cyanobacteria is to contribute to the neutralization of derivative toxics of oxygen (ROS).

Beside carotenoids, many photoprotective substances play a key role in preventing UV-induced direct or indirect damage to essential biomolecules (Karsten 2008). The most common photoprotective sunscreens are mycosporines-like amino acids (MAAs). MAAs act as photo-protective UV filters and/or as antioxidants. According to (Lee and Shiu 2009), MAAs are able to effectively dissipate absorbed radiation without producing reactive oxygen species. MAAs were firstly discovered in fungi species but are in fact largely distributed in marine, freshwater or aeroterrestrial algae and cyanobacteria (Sinha et al. 2007). Other photoprotective substances are scytonemins, which have been reported in all cyanobacteria groups and which are detected in high radiation habitats. Phenolic compounds such as phlorotannins are exclusively found in brown macroalgae. While scytonemin acts as a UV-A sunscreen, phenolic compounds absorb mainly UV-C and partly UV-B radiations.

If one now considers the repair mechanisms, all photosynthetic organisms possess strategies to repair a large part of the damage occurring on DNA (photoreactivation and base excision repair) and/or on UV-damaged protein by *de novo* biosynthesis (e.g. D1 and D2 reaction center subunits of PSII) (Hollósy 2002; Jansen et al. 1998; Karsten 2008).

4. Research Project: The Use of UV-C Irradiation to Eradicate Biofouling by Algae

This study aims to clarify the effect of UV-C irradiation on algal species growing on building facades, historical monuments and subterranean environments in order to develop a new method of treatment against algal contamination. To develop an accurate eradication process, it is imperative to well understand the physiology of the algae implied in biofouling and the effect of UV-C irradiation on them. Several methodological approaches are used, which are more detailed in the following paragraphs. Some physiological parameters and cell proliferation are studied to understand the effect of ultraviolet light at the cellular, molecular and at the biofilm scale. For this purpose, two methods of culture were developed: one in liquid medium and one on solid support. In addition, an example of the application of this method in natural conditions is given to highlight the efficiency of the treatment.

Table 1. Effects, damages and responses of living organisms exposed to different types of UV (A, B or C)

Types of UV	Biological models	Effects/Damages/Responses	Authors/dates
UV-A	Red algae (*Chondrus crispus*)	-decrease of quantum yield -decrease of pigment contents	(Yakovleva and Titlyanov 2001)
UV-B	Macro-algae and cyanobacteria	-photosynthesis inhibition -damages on nucleus (phototransformation resulting in the production of cyclobutane pyrimidine dimmers, CPDs) -synthesis of sunscreens compounds (Mycosporines-like amino-acids, MAAs) -Reactive Oxygen Species formation	(Karsten 2008) (Sinha and Häder 2002) (Ma and Gao 2010) (Hsu et al. 1998)
	Bacteria (*Deinococcus radiodurans*, *Pseudomonas fluorescens*, *Serratia marcescens*, and *Staphylococcus epidermidis*)	-decrease between 50 and 100% of the viability for UV exposures in seconds	(Menetrez et al. 2010)
	Litter fungi (*Aspergillus fumigatus*, *Penicillium hordei*, *P. janczewskii*, *P. spinulosum* and *P. Purpurogenum*)	-reductions in mycelia extension rate and spore germination	(Moody et al. 1999)
UV-C	Higher plant (Aquatic fern, *Azolla microphylla*)	-decrease of biomass, photo bleaching of chlorophyll, damage to the nitrogen fixation and inhibition of PSII electron transport	(Jayakumar et al. 1999)
	Fungi (*Aspergillus flavus*, *Aspergillus niger*, *Penicillium corylophilum* and *Eurotium rubrum*)	-decrease of spores viability -photo protective component melanin increases the survival and longevity of fungal spores	(Begum et al. 2009)
	Unicellular green alga (*Chlamydomonas reinhardtii*)	-induction of Programmed Cell Death: cellular morphological changes and DNA fragmentation	(Moharikar et al. 2006)

4.1. Algae, Culture Conditions and UV-C Exposure

The algae used in our study were all collected from natural outdoor sites. As a consequence, it doesn't matter if it's a single strain but rather of a natural mixture of genera growing together.

The first algal mixture was harvested in three caves of South West of France (Domme, Laugerie-Haute, Combarelles, Department of Dordogne, France). It is composed of *Klebsormidium* sp. (Klebsormidiales, Charophyceae) and *Chlorella* sp. (Chlorellaceae, Chlorophyta). It was used for the algal culture on solid support.

The second algal mixture was harvested in the Moidon's caves (Department of Jura, France). It turned out to be a mixture of two genera, *Bracteacoccus sp.* and *Chlorella sp.*, belonging in the Chlorophyta division to the Chlorococcales and to the Chlorellaceae family. These algal strains have been used for algal culture in liquid medium.

The growth of the algal cells was performed in a complete culture medium especially adapted to algal development (BG11 freshwater solution, C3061, SIGMA) to minimize any contamination by other organisms. The erlenmeyers were permanently horizontally agitated at 100 rpm for a proper oxygenation of the medium and exposed to a photoperiod of 16h at 130µmol photons.m^{-2}.s^{-1} and 8h in the dark with a thermoperiod of 24°C and 18°C respectively.

In order to mimic the natural conditions in caves, the algae were inoculated on limestone blocks of 6cm in length, 4cm in width and 6mm in thickness provided by the LRMH. A standardized protocol of algae inoculating has been elaborated to homogenize the number of algal cells deposited on the blocks (Figure 1).

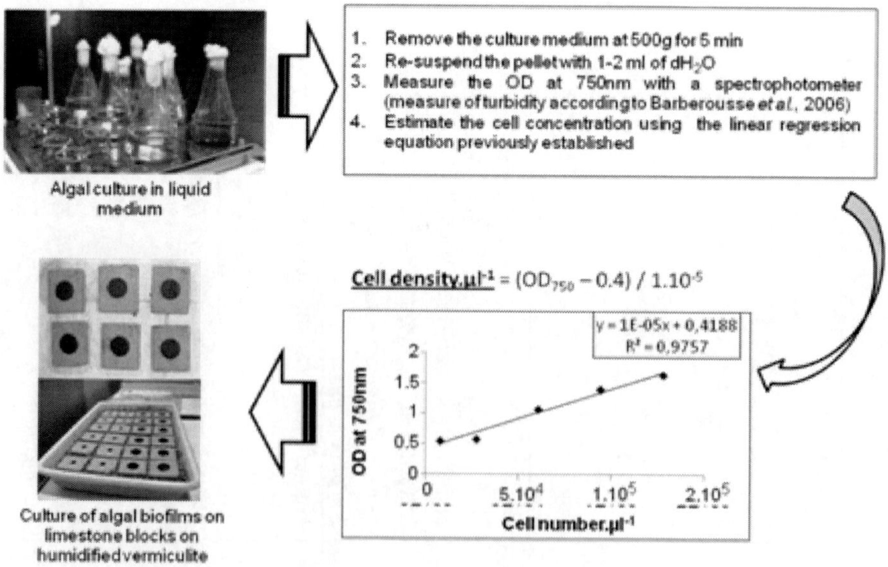

Figure 1. Standardized protocol of algal inoculation on limestone blocks.

UV-C treatments are performed with an "UV-C Box" (Figure 2) containing four UV-C lamps of 25W (TUV 25W G25T8, PHILIPS). Algal cells cultivated in liquid medium are exposed to UV-C in Petri dishes without lids approximately 30cm from the UV-C lamps.

Algal biofilms on limestone blocks are deposited on moist vermiculite and exposed to UV-C in the same conditions. All UV-C treatments are applied intermittently with period of darkness to avoid an excessive temperature rise. One hour of UV-C is always followed by fifteen minutes of darkness. In the case of algal biofilms deposed on limestone blocks, an additional treatment has been used. To test a photo-inhibiting effect on algal biofilms, algae were exposed to High Light (HL: 6 hours corresponding to 1380 µmol m^{-2} s^{-1}) provided by CL 1500 ECO SCHOTT (150 W) in addition to the previously described UV-C treatment.

For all experiments in liquid medium, a cellular suspension at approximately 1.10^8 cells.ml^{-1} has been used.

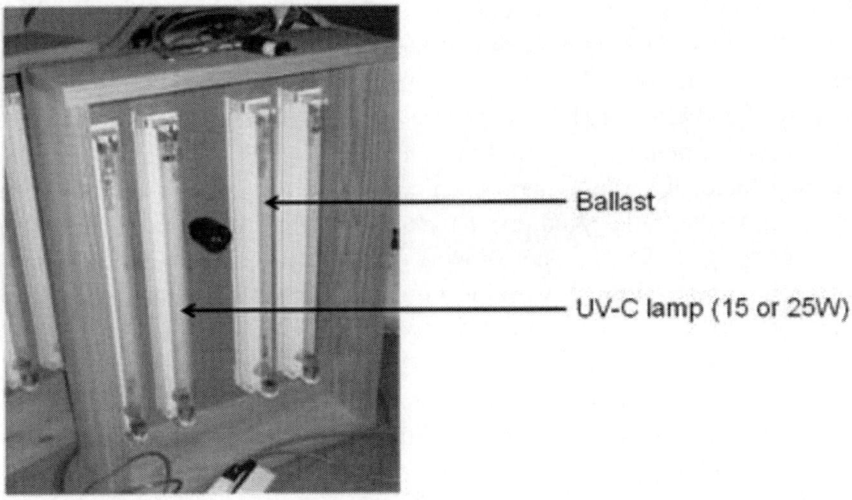

Figure 2. UV-C Box (maximal power: 100W; wavelength of lamps: 253.4nm).

During exposure, UV-C measurements have been realized with a photoradiometer (HD2102.2, DELTA OHM) using a UV-C probe (LP 471UVC). The measure is given in kJ.m^{-2} or in W.m^{-2}.

4.2. Experiments and Obtained Results

4.2.1. On Algal Suspension

Physiological Parameters

As shown in the part 3.2, ultraviolet radiation causes deleterious effects on vital metabolic functions of photosynthetic organisms such as algae, even if most of the studies reported results obtained with UV-A or UV-B and if data concerning UV-C irradiation remained scarce. From these reports, it appears that the most severely affected parameters are the viability of cells, the photosynthetic activity and the pigment contents.

In a first time in order to highlight the impact of UV-C radiation on our algal culture, we chose to follow physiological parameters that, according to most authors, are obviously affected by UV-C exposure and at the same time easy to determine. These parameters are the cellular viability, the photosynthetic activity and the pigment contents (chlorophyll a, b and carotenoids).

The cellular viability is measured by a colorimetric method with Trypan blue. Depending on whether the cell is dead or alive, the dye is released, or not, out of the cell. Living cells thus appear uncolored whereas dead cells present a blue coloration. This method thus authorizes the discrimination between dead or alive cells. The cells' counting is performed under a microscope using a Malassez cell and the percentage of living cells is calculated.

The photosynthetic activity is measured by a polarometric method using an oxygraph (Oxygraph System, Hansatech), which measures O_2 released through photosynthesis in the algal suspension. Finally, the pigment contents are measured using a spectrophotometric method after methanol extraction (according to (Devesa et al. 2007).

The obtained results show that UV-C irradiation leads to significant changes in the pigment contents. The chlorophyll a and chlorophyll b contents were significantly affected after respectively 1 h and 2.5 h of UV-C exposure ($p<0.01$). Indeed, after a transient increase, the chlorophyll a and b contents recovered their initial values at respectively 2.5h and 5h of UV-C exposure time and, at least in regard to chlorophyll a, a significant decrease is observed at 5h of UV-C exposure time. And, if the exposure time is prolonged, a decrease in chlorophyll b is also registered (data not shown). These results are similar to those obtained by (Yakovleva and Titlyanov 2001) working on the red alga *Chondrus crispus* exposed to UV-A and UV-B radiation except that the UV doses that we have applied were about 3 time stronger (12.4 W.m-2 in our work vs. 3.6 W.m-2 in their experiment) and that the shorter wavelength that we have used are known to be much more harmful than UVB.

In fact, study concerning the effects of UV-C on photosynthetic pigments are very scarce, and the most available data concerns UV-B and/or UV-A. We must therefore refer to results obtained with UV-B, bearing in mind that the effects of UV-C are exacerbated compared to UV-B, because of their higher energy. With UV-B irradiation, it has been reported that a destruction of photosynthetic pigments, especially chlorophyll a, may occur at least with respect to higher plants (Lidon and Ramalho 2011; Teramura 1983) but also on algae (Hupel et al. 2011). Further work, using UV exposures at doses much lower than we used, showed on the contrary no effects on macroalgae (Roleda et al. 2010). As reported by (Bischof et al. 2007), chlorophyll a seems to be more sensitive to UV radiation than chlorophyll b and the chlorophyll a/b ratio, as in our work, is often decreased (Hupel et al. 2011).

Table 2. Pigment contents and chlorophyll a/b ratio in control and UV treated samples at different doses of UV-C

Pigment contents ($\mu g.ml^{-1}$)	Control	UV treated samples (1h)	UV treated samples (2.5h)	UV treated samples (5h)
Chl a	4.19 ± 0.17	6.35 ± 0.10*	4.7 ± 0.59	1.61 ± 0.22*
Chl b	2.06 ± 0.17	2.5 ± 0.26	3.24 ± 0.13*	1.94 ± 0.19
Chl a/b ratio	2.03 ± 0.09	2.55 ± 0.22*	1.45 ± 0.16*	0.83 ± 0.03*
Carotenoids	2.09 ± 0.13	1.59 ± 0.08*	0.72 ± 0.2*	0.16 ± 0.01*

Means values ± standard errors. Asterisks indicate significant differences compared to the control samples (ANOVA, $p<0.01$). Approximate UV-C doses in $kJ.m^{-2}$: 1h or 29.7 $kJ.m^{-2}$, 2.5h or 74.3 $kJ.m^{-2}$, 5h or 149 $kJ.m^{-2}$.

Concerning the carotenoid contents, our results show that UV-C exposure causes a significant decrease even after only 1h of UV-C exposure time. Our results are consistent with those of (Hupel et al. 2011) working on macroalgae. On the contrary, (Yakovleva and Titlyanov 2001), obtained under UV-A or UV-B exposures, an accumulation of carotenoids which is supposed to counteract the oxidative stress induced by UV radiation (Prasad et al. 2011; Stahl and Sies 2003). Other authors working with quite low UV A and/or B doses detected no effect on the carotenoid level (Roleda et al. 2010).

In order to ascertain that UV-C exposure effectively decreases the algal cell proliferation, controlled and exposed samples have been re-inoculated in a fresh medium and allowed to grow under optimal conditions.

Two weeks after re-inoculation, control samples have shown a good physiological state with an exponential growth during this period, while samples exposed at 180 minutes to UV-C did not show any cell proliferation (Borderie et al. 2011). These observations have been confirmed by establishing, through the counting, the ratio between dead and living cells (Figure 3).

Indeed, this graph shows that cellular viability is severely decreased after different exposure times to the UV-C. When algal cells were exposed to UV-C for 60 minutes, the viability is decreased from about 40% compared to the control, and after a 180 minutes of exposure, only a few surviving cells could be detected. According to the literature, UV-induced cell death frequently results from damage caused to DNA (Danon and Gallois 1998; Hollósy 2002; Moharikar et al. 2006). Under UV-B exposure pyrimidine dimmers formation was reported (Strid et al. 1994). But other cell components might as well have been affected, like endomembranes, which integrity is affected under UV exposure (Meindl and Lütz 1996) and proteins (Hollósy 2002; Jansen et al. 1998), and also due to the deleterious effects on cellular membrane integrity and on proteins as well as on the photosynthetic apparatus through its pigments, as we have shown previously.

The measure of the net photosynthetic rate has been undertaken on the same samples, and is also presented in Figure 3. The obtained results show that the net photosynthetic rate is more precociously and more severely affected by UV-C exposure than cellular viability, because a more than 50% inhibition is observed after only 20 minutes of UV-C exposure. Moreover, after 1h of UV-C treatment, the inhibition reaches 100%.

This net photosynthetic rate collapse may be explained by two complementary factors, *i.e.* a decrease in the photosynthetic activity and an increase of the cells respiration rate.

The photosynthetic apparatus is in fact a prime target from UV radiations. As we have shown previously, the pigment content is affected by UV-C. But it was also shown that photosynthesis is mainly affected by an accelerated turnover and an increased destruction of one of the major components of PS II, the D1 protein (Hollósy 2002; Karsten 2008; Ma and Gao 2010).

Furthermore, damage to thylakoid membranes (increased permeability) and reduced RubisCO activity were also reported (Karsten 2008). Concerning the O_2 consumption *i.e.* the respiration, it seems likely to be increased given the strong need of energy required by cells which are undertaking various process of repair and *de novo* synthesis. Among the *de novo* synthesis, one can cite protective components like various flavonoids and MAAs. An increased need of energy is also required from repair processes including DNA repair and other costly pathways.

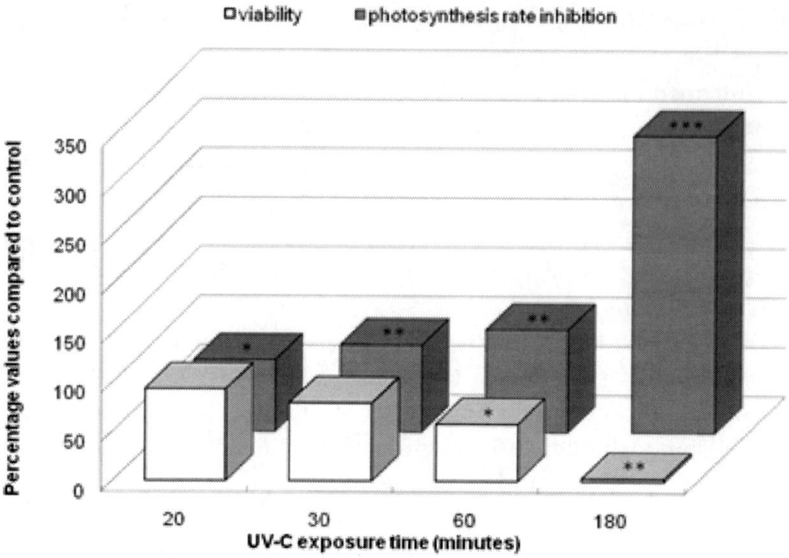

Figure 3. Viability and photosynthesis rate inhibition of algal cells according to the UV-C exposure time. Values are means (n=3). Stars represent significant differences with Anova test compared to control sample at p<0.05 for *, p<0.01 for ** and p<0.001 for ***.

Molecular Parameters

The effect of UV-C on the integrity of nuclear apparatus was investigated on the same algal strains in order to detect a possible Programmed Cell Death (PCD) of algal cells exposed to UV-C.

Firstly, DNA fragmentation has been measured with a flow cytometer (CXP Cytometer, BECKMAN COULTER) using a TUNEL kit (*In situ* Cell Death Detection Kit, Fluorescein, ROCHE). TUNEL, (for *Terminal deoxynucleotidyl transferase-mediated dUTP nick end-labeling*), is a method to detect DNA strand breaks ("nicks") by labeling free 3'-OH termini by enzymatic reaction. Previous to the TUNEL labeling, cells have been fixed with freshly prepared 4% paraformaldehyde and permeabilized with 0.1% TRITON X-100 and 0.1% citrate sodium. DNA fragmentation has also been detected by fluorescence microscopy with a fluorescent marker (Hoechst 32258, SIGMA ALDRICH) to visualize chromatin condensation and thus DNA fragmentation.

Obtained results are presented in Figure 4 with the percentage of TUNEL positive cells (measure on 20,000 cells for each treatment) and pictures indicating cellular nucleus damages.

Figure 4 shows that control cells as well as 1h UV-C exposed cells present no significant cell death rate, with a normal cell death around 5%. After 10h of UV-C treatment, results show that TUNEL positive cells *i.e.* dead or heavily damaged cells are around 35% (p<0.05 compared to control). These results obtained by flow cytometry are confirmed by the observations under fluorescence microscopy of the nucleus of the cells. Control cells show a normal and regular nucleus with uniformly distributed chromatin, whereas 10h of UV-C exposed cells show fragments of nucleus apparatus randomly distributed in the cell.

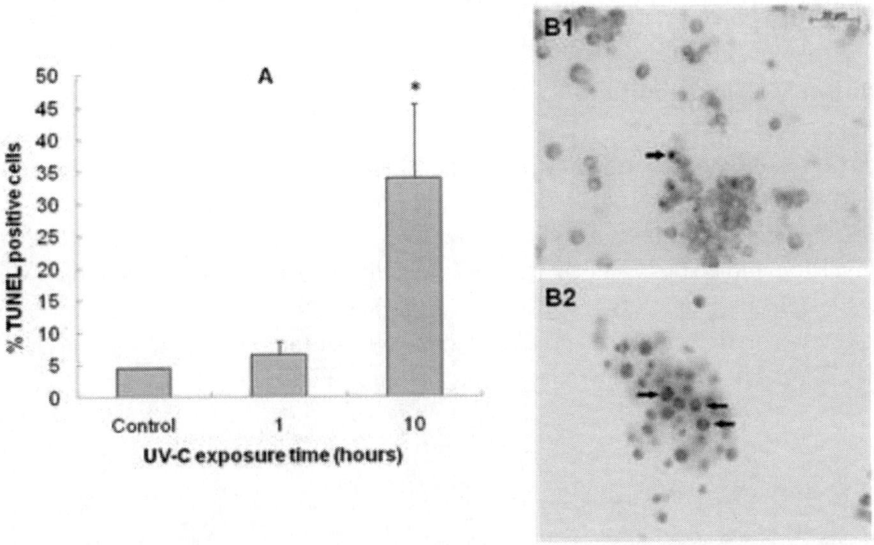

Figure 4. Percentage of TUNEL positive cells according to the UV-C exposure time (**A**). Values are means (n=3) ± standard errors obtained by flow cytometry. Star represents significant difference at $p<0.05$ with ANOVA test. Visualization of the chromatin condensation by fluorescence microscopy after Hoechst dying (**B**). **B1**: control cells; **B2**: exposed cells at 10h of UV-C. Arrows indicate controlled and exposed cellular nucleus.

These results also confirmed the effect of UV-C irradiation at the DNA level from the algal cells exposed. The UV radiations are known to cause damage on nucleic acids, to disrupt transcription and replication processes during cellular division and finally to induce cell death (Karsten 2008). One form of cell death generated by UV-irradiation, and mainly by UV-C, is the so-called Programmed cell Death (PCD) and particularly the "apoptosis" phenomenon. Death by apoptosis differs from death by necrosis in the way that necrosis is not, contrary to apoptosis, a genetically-controlled process. Moreover, necrosis affects a group of contiguous cells, whereas apoptosis affects individual cells.

Apoptosis is a genetically-controlled process occurring in a cell and characterized by chromatin condensation, nucleus and cell fragmentation and cellular morphological changes. The DNA fragmentation mostly observed in the case of apoptosis is due to a caspases-activated DNase (Danon and Gallois 1998).

Few studies have shown that UV-C exposure can induce apoptosis in photosynthetic organisms. The first, (Danon and Gallois 1998) highlight the phenomenon on *Arabidopsis thaliana*. Then, (Moharikar et al. 2006) working on an unicellular Chlorophyta, *Chlamydomonas reinhardtii*, showed that it may occur in algae too. More recently, (Gao et al. 2008) working on *A. thaliana* demonstrate that ROS play a crucial role in the cell death process. Yet it has long been established that irradiation by UV-B induces the formation (Hideg and Vass 1996). And it was even show that under UV-B exposure, the superoxide ion $O_2^{\bullet -}$ was preferentially produced, indicating a dysfunction in PSII. This supports the observation previously reported that the PSII structure may be one of the first targets of UV radiation. Our results constitute, then, additional proof that UV-C irradiation causes PCD on unicellular algae.

4.2.2. On Algal Biofilm

Proliferation Measurements

As described in section 4.1, algal cells were inoculated from concentrated algal suspension on solid support to form biofilm and so mimic the growth conditions in cavese. Covering rate measurements were conducted using a microscope (Stemi 2000-CS ZEISS) connected to a camera to take pictures of samples. Pictures are thresholded after black and white converting and covering rates are automatically measured as a percentage using ImageJ software. For all treatments (15h, 2x15h and High Light treatments), samples were measured just after the inoculation (day 0), and during one week at the rate of one measurement per day (day 1, day 2 ...). The Obtained results are presented in Figure 5.

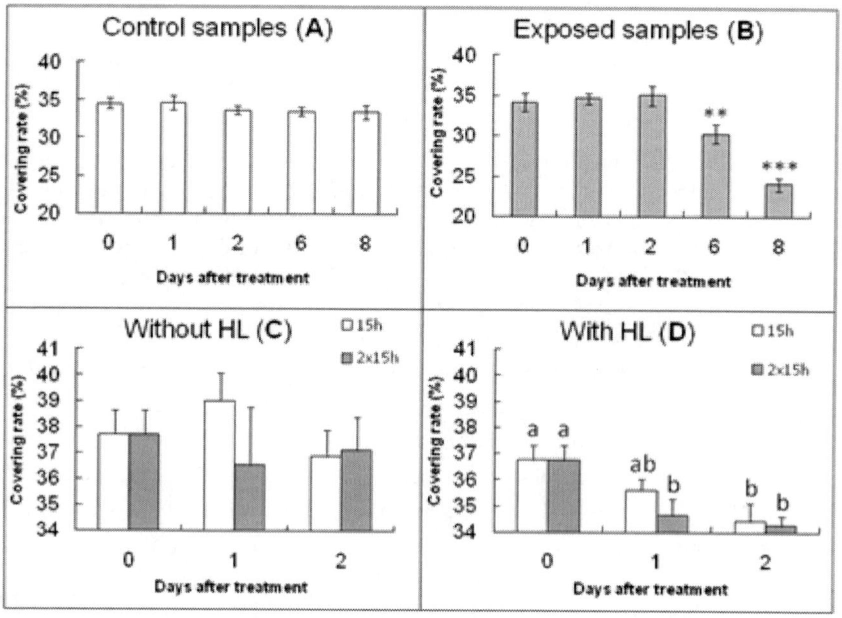

Figure 5. Daily measurements of algal covering rate after treatment of 15h of UV-C for (**B**) and 2 times 15h (2x15h) for (**C**) and (**D**). For exposed samples, UV-C treatment is applied just after the initial measure at day 0. HL: High Light treatment (6h). Values are means (n=3) ± standard errors. Stars represent significant differences with TukeyHSD test at p<0.01 for ** and p<0.001 for ***. In the graph (**D**), different letters indicate significant differences at p<0.05 with TukeyHSD test.

A preliminary 3 hour exposure to UV-C was sufficient to reduce from 50 % the covering rate after one week growth (Borderie et al. 2011), at least in the case of a thin biofilm. Moreover, a 9h treatment with UV-C stops the proliferation of a thick biofilm however without reducing its surface at the end of the week. In the same way, always in the case of a thick biofilm, a 9h UV-C exposure followed by a 6 h high light treatment conduce a reduction of about 10% of the covering rate after one week of growth (p<0.05). In Figure 5, thick algal biofilms were submitted to a single exposure phase of 15h (A and B) or to two consecutive UV-C exposure phases of 15h with (D) or without (C) subsequent high light treatment (HL). We can observe that the most efficient treatment consists of two exposure phases of 15h UV-C followed by HL treatment. In fact, only one day after the exposition, the algal biofilm was significantly reduced compared to the control sample. The time required to achieve a similar result with a single 15h exposure phase coupled with HL treatment was 2 days. And without HL treatment, a significant decrease of the biofilm occurred only after one week in the case of a single exposure (B) and after 4 days in the case of two consecutive exposures (data not shown). They were no data available in the literature concerning the effects of UV-C treatment on algal biofilm inoculated on a solid support. Our results show that UV-C treatment(s) is efficient to limit or reduce algal proliferation on a solid support. Moreover, a high light treatment consecutive to the UV-C exposure enhances the treatment efficiency, due to photo inhibition (Altamirano et al. 2004; Hanelt et al. 2006). After a long-term treatment (2x15h of UV-C) the covering rate was decreased, which indicates a bleaching of the algal cells in the biofilm. That bleaching is due to a chlorophyll degradation inevitably followed by a decrease in cell viability. However, in that experiment, some biochemical parameters were followed concomitantly to the algal covering rate. So, the determination of the pigment content and of the cell viability revealed that on the same spot where we detected a decreased cell covering, the treatment had no effect on these parameters. The explanation certainly lies in the biofilm thickness. The underlying cells probably benefit from a protective effect from the overlying cells that might filter the UV-C radiation. So, according to the first published results (Borderie et al. 2011), it appears that the thickness of an algal biofilm may strongly condition the efficiency of the UV treatment. It must also be underlined that the effect of High Light treatment after an UV-C exposure causes more damage than with UV-C exposure only.

In-Situ Preliminary Test

In May 2010, the LRMH conducted a preliminary test of UV-C treatment on a wall contaminated by green algae. The UV-irradiation was performed in the Abbey of Saint-Martin in the Laon city (Department of Aisne, France). A treatment consisting of 16 hours of UV-C exposure at 30W was applied during the night. Visual observations of this first test are presented in Figure 6. On the first picture (6A), before the treatment, the algal biofilm covers an important part of the wall. After the treatment (6B) with 16 hours of UV-C, the algal biofilm was strongly affected, showing a total discoloration and having completely disappeared.

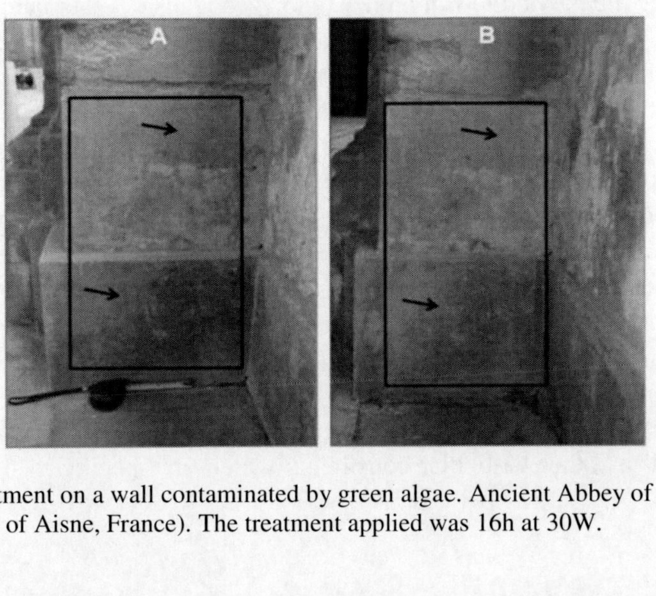

Figure 6. UV-C treatment on a wall contaminated by green algae. Ancient Abbey of Saint-Martin, city of Laon (department of Aisne, France). The treatment applied was 16h at 30W.

CONCLUSION

Obviously, algal proliferation on monument and art works constitute a widespread difficulty that has to be overcome with a view to leave an intact architectural and cultural heritage for future generations. Colonization by algae is more than just an aesthetic nuisance through the stains they cause; they are also a threat to the durability of the support they occupy because of their biochemical interactions with it. Up to now, the methods used to clean the colonized support and get rid of the algae were mechanical (brushing) or chemical with the use of detergents and herbicides. Taking into account the fragility of some areas where they are supposed to be used, it appears nowadays urgent to develop softer methods, allowing the preservation of the environment and avoiding the spread of pollution or any risk for other living organisms. The UV-C irradiation investigated in this present work may constitute a possible efficient and without risk alternative. This innovative technique that we have developed in our work and which is already used in other industries for the eradication of microorganisms is an emerging technique in the field of heritage conservation. The works on the development of this technique are made difficult because of the combination of several factors that are:

(i) the limited knowledge available regarding the biochemistry of unicellular or filamentous algae that colonize facades and monuments
(ii) the limited data on the physiological effects of UV-C on this type of organism,
(iii) the resistance acquired by these algae through living in extreme conditions of sunlight exposure and water shortage.

Nevertheless, results we have already obtained are very encouraging. These works have firstly allowed some progress in the comprehension of the physiological responses of algae submitted to UV-C exposure. The efficiency of the UV-C exposure method to limit the algal proliferation and even to eradicate them was demonstrated on algal suspension as well as on algae inoculated on limestone blocks, a growing condition very close to that really encountered on monuments, caves and other areas. Moreover, very promising results have been obtained using the method *in situ* on real colonized walls or other natural supports.

In the future, the UV-C exposure technique must of course being perfected. Firstly, it seems extremely important to deepen the study of the effects of the UV-C treatment on physiological, biochemical and molecular parameters of algal cells. A better understanding of the types of cellular damages caused and of the types of protection that may be set up, could indeed allow us to select the irradiation parameters (doses, UV-C combined or not with a saturating light, alternation between exposure and rest period ...) the most deleterious to microorganisms and therefore the most effective for their eradication. Special attention will be paid to the possible occurrence of MAAs, which are UV-screen compounds potentially synthesized by some unicellular algae and Cyanobacteria under UV exposure.

At the same time, we plan to examine the time required for a re-colonization or re-proliferation after an UV treatment in order to establish the best treatment protocol to apply, taking into account the type of microorganisms present, the thickness and the overall size of the biofilm as well as other particular constraints which are specific to the treated site (for example the schedule of visits). In this spirit, many experiments will be conducted *in situ* on monuments or natural areas colonized by algae.

REFERENCES

Altamirano, M., Murakami, A., Kawai, H., 2004. High light stress in the kelp Ecklonia cava. *Aquatic Botany* 79, 125-135.

Alum, A., Rashid, A., Mobasher, B., Abbaszadegan, M., 2008. Cement-based biocide coatings for controlling algal growth in water distribution canals. *Cement and Concrete Composites* 30, 839-847.

Barberousse, H., Ruot, B., Yéprémian, C., Boulon, G., 2007. An assessment of façade coatings against colonisation by aerial algae and cyanobacteria. *Building and Environment* 42, 2555-2561.

Barberousse, H., Tell, G., Yéprémian, C., Couté, A., 2006. Diversity of algae and cyanobacteria growing on building facades in France. *Algological Studies/Archiv für Hydrobiologie*, Supplement Volumes 120, 81-105.

Begum, M., Hocking, A.D., Miskelly, D., 2009. Inactivation of food spoilage fungi by ultra violet (UVC) irradiation. *International Journal of Food Microbiology* 129, 74-77.

Bischof, K., Gómez, I., Molis, M., Hanelt, D., Karsten, U., Lüder, U., Roleda, M.Y., Zacher, K., Wiencke, C., 2007. Ultraviolet radiation shapes seaweed communities. *Protoplasma* 243, 105-116.

Blokhina, O., Virolainen, E., Fagerstedt, K.V., 2003. Antioxidants, Oxidative Damage and Oxygen Deprivation Stress: a Review. *Annals of Botany* 91, 179-194.

Borderie, F., Laurence, A.-S., Naoufal, R., Faisl, B., Geneviève, O., Dominique, R., Badr, A.-S., 2011. UV-C irradiation as a tool to eradicate algae in caves. *International Biodeterioration & Biodegradation* 65, 579-584.

Danon, A., Gallois, P., 1998. UV-C radiation induces apoptotic-like changes in Arabidopsis thaliana. *FEBS Letters* 437, 131-136.

De Muynck, W., Ramirez, A.M., De Belie, N., Verstraete, W., 2009. Evaluation of strategies to prevent algal fouling on white architectural and cellular concrete. *International Biodeterioration & Biodegradation* 63, 679-689.

Devesa, R., Moldes, A., Díaz-Fierros, F., Barral, M.T., 2007. Extraction study of algal pigments in river bed sediments by applying factorial designs. *Talanta* 72, 1546-1551.

Feng, H., Li, S., Xue, L., An, L., Wang, X., 2007. The interactive effects of enhanced UV-B radiation and soil drought on spring wheat. *South African Journal of Botany* 73, 429-434.

Gao, C., Xing, D., Li, L., Zhang, L., 2008. Implication of reactive oxygen species and mitochondrial dysfunction in the early stages of plant programmed cell death induced by ultraviolet-C overexposure. *Planta* 227, 755-767.

Gaylarde, P.M., Gaylarde, C.C., 2000. Algae and cyanobacteria on painted buildings in Latin America. *International Biodeterioration & Biodegradation* 46, 93-97.

Grobbelaar, J.U., 2000. Lithophytic algae: A major threat to the karst formation of show caves. *Journal of Applied Phycology* 12, 309-315.

Grütter, M.G., 2000. Caspases: key players in programmed cell death. *Current Opinion in Structural Biology* 10, 649-655.

Guillitte, O., 1995. Bioreceptivity: a new concept for building ecology studies. *Science of The Total Environment* 167, 215-220.

Gómez, I., Pérez-Rodríguez, E., Viñegla, B., Figueroa, F.L., Karsten, U., 1998. Effects of solar radiation on photosynthesis, UV-absorbing compounds and enzyme activities of the green alga Dasycladus vermicularis from southern Spain. *Journal of Photochemistry and Photobiology B: Biology* 47, 46-57.

Hanelt, D., Hawes, I., Rae, R., 2006. Reduction of UV-B radiation causes an enhancement of photoinhibition in high light stressed aquatic plants from New Zealand lakes. *Journal of Photochemistry and Photobiology B: Biology* 84, 89-102.

Hideg, É., Vass, I., 1996. UV-B induced free radical production in plant leaves and isolated thylakoid membranes. *Plant Science* 115, 251-260.

Hodoki, Y., 2005. Effects of solar ultraviolet radiation on the periphyton community in lotic systems: comparison of attached algae and bacteria during their development. *Hydrobiologia* 534, 193-204.

Hollósy, F., 2002. Effects of ultraviolet radiation on plant cells. *Micron* 33, 179-197.

Hsu, T., Ho, J.-C., Chao, C.-C., 1998. Purification of a UV-damaged-DNA binding activity from cell-free extracts of unicellular alga Chlorella pyrenoidosa. *Plant Science* 138, 137-147.

Hupel, M., Lecointre, C., Meudec, A., Poupart, N., Gall, E.A., 2011. Comparison of photoprotective responses to UV radiation in the brown seaweed Pelvetia canaliculata and the marine angiosperm Salicornia ramosissima. *Journal of Experimental Marine Biology and Ecology* 401, 36-47.

Janknegt, P.J., Rijstenbil, J.W., van de Poll, W.H., Gechev, T.S., Buma, A.G.J., 2007. A comparison of quantitative and qualitative superoxide dismutase assays for application to low temperature microalgae. *Journal of Photochemistry and Photobiology B: Biology* 87, 218-226.

Jansen, M.A.K., Gaba, V., Greenberg, B.M., 1998. Higher plants and UV-B radiation: balancing damage, repair and acclimation. *Trends in Plant Science* 3, 131-135.

Jayakumar, M., Eyini, M., Selvinthangadurai, P., Lingakumar, K., Premkumar, A., Kulandaivelu, G., 1999. Changes in Pigment Composition and Photosynthetic Activity of Aquatic Fern (<i>Azolla Microphylla Kaulf.) Exposed to Low Doses of UV-C (254 nm) *Radiation. Photosynthetica* 37, 33-38.

Jenkins, G.I., 2009. Signal Transduction in Responses to UV-B Radiation. *Annual Review of Plant Biology* 60, 407-431.

Karsten, U., 2008. Defense Strategies of Algae and Cyanobacteria Against Solar Ultraviolet Radiation, In Algal Chemical Ecology. ed. C.D. Amsler, pp. 273-296. Springer Berlin Heidelberg.

Karsten, U., Lembcke, S., Schumann, R., 2007. The effects of ultraviolet radiation on photosynthetic performance, growth and sunscreen compounds in aeroterrestrial biofilm algae isolated from building facades. *Planta* 225, 991-1000.

Lee, T.-M., Shiu, C.-T., 2009. Implications of mycosporine-like amino acid and antioxidant defenses in UV-B radiation tolerance for the algae species Ptercladiella capillacea and Gelidium amansii. *Marine Environmental Research* 67, 8-16.

Lidon, F.C., Ramalho, J.C., 2011. Impact of UV-B irradiation on photosynthetic performance and chloroplast membrane components in Oryza sativa L. *Journal of Photochemistry and Photobiology B: Biology* 104, 457-466.

Ma, Z., Gao, K., 2010. Spiral breakage and photoinhibition of Arthrospira platensis (Cyanophyta) caused by accumulation of reactive oxygen species under solar radiation. *Environmental and Experimental Botany* 68, 208-213.

Mahdavian, K., Ghorbanli, M., Kalantari, K., 2008. Role of salicylic acid in regulating ultraviolet radiation-induced oxidative stress in pepper leaves. *Russian Journal of Plant Physiology* 55, 560-563.

Maxwell, K., Johnson, G.N., 2000. Chlorophyll fluorescence—a practical guide. *Journal of Experimental Botany* 51, 659-668.

Meindl, U., Lütz, C., 1996. Effects of UV irradiation on cell development and ultrastructure of the green alga Micrasterias. *Journal of Photochemistry and Photobiology B: Biology* 36, 285-292.

Menetrez, M.Y., Foarde, K.K., Dean, T.R., Betancourt, D.A., 2010. The effectiveness of UV irradiation on vegetative bacteria and fungi surface contamination. *Chemical Engineering Journal* 157, 443-450.

Moharikar, S., D'Souza, J.S., Kulkarni, A.B., Rao, B.J., 2006. Apoptotic-Like Cell Death Pathway Is Induced In Unicellular Chlorophyte Chlamydomonas Reinhardtii (Chlorophyceae) Cells Following Uv Irradiation: Detection And Functional Analyses1. *Journal of Phycology* 42, 423-433.

Moody, S.A., Newsham, K.K., Ayres, P.G., Paul, N.D., 1999. Variation in the responses of litter and phylloplane fungi to UV B radiation (290–315 nm). *Mycological Research* 103, 1469-1477.

Moreau, C., Vergès-Belmin, V., Leroux, L., Orial, G., Fronteau, G., Barbin, V., Water-repellent and biocide treatments: Assessment of the potential combinations. *Journal of Cultural Heritage* 9, 394-400.

Moreau, C., Vergès-Belmin, V., Leroux, L., Orial, G., Fronteau, G., Barbin, V., 2008. Water-repellent and biocide treatments: Assessment of the potential combinations. *Journal of Cultural Heritage* 9, 394-400.

Najeeb, U., Xu, L., Ahmed, Z., Rasheed, M., Jilani, G., Naeem, M., Shen, W., Zhou, W., 2011. Ultraviolet-C mediated physiological and ultrastructural alterations in <i>Juncus effusus L. shoots. *Acta Physiologiae Plantarum* 33, 481-488.

Nugari, M.P., Pietrini, A.M., Caneva, G., Imperi, F., Visca, P., 2009. Biodeterioration of mural paintings in a rocky habitat: The Crypt of the Original Sin (Matera, Italy). *International Biodeterioration & Biodegradation* 63, 705-711.

Prasad, K.N., Chew, L.Y., Khoo, H.E., Yang, B., Azlan, A., Ismail, A., 2011. Carotenoids and antioxidant capacities from Canarium odontophyllum Miq. fruit. *Food Chemistry* 124, 1549-1555.

Rindi, F., Guiry, M.D., López-Bautista, J.M., 2008. Distribution, Morphology, And Phylogeny Of Klebsormidium (Klebsormidiales, Charophyceae) In Urban Environments In Europe1. *Journal of Phycology* 44, 1529-1540.

Roleda, M., Lütz-Meindl, U., Wiencke, C., Lütz, C., 2010b. Physiological, biochemical, and ultrastructural responses of the green macroalga Urospora penicilliformis from Arctic Spitsbergen to UV radiation. Protoplasma 243, 105-116.

Rozema, J., Björn, L.O., Bornman, J.F., Gaberscik, A., Häder, D.P., Trost, T., Germ, M., Klisch, M., Gröniger, A., Sinha, R.P., Lebert, M., He, Y.Y., Buffoni-Hall, R., de Bakker, N.V.J., van de Staaij, J., Meijkamp, B.B., 2002. The role of UV-B radiation in aquatic and terrestrial ecosystems--an experimental and functional analysis of the evolution of UV-absorbing compounds. *Journal of Photochemistry and Photobiology B: Biology* 66, 2-12.

Ryan, K.G., 1992. UV radiation and photosynthetic production in Antarctic sea ice microalgae. *Journal of Photochemistry and Photobiology B: Biology* 13, 235-240.

Sarró, M.I., García, A.M., Rivalta, V.M., Moreno, D.A., Arroyo, I., 2006. Biodeterioration of the Lions Fountain at the Alhambra Palace, Granada (Spain). *Building and Environment* 41, 1811-1820.

Scheerer, S., OrtegaMorales, O., Gaylarde, C., 2009. Chapter 5 Microbial Deterioration of Stone Monuments--An Updated Overview, In Advances in Applied Microbiology. ed. S.S.a.G.M.G. Allen I. Laskin, pp. 97-139. Academic Press.

Shao, H.-b., Chu, L.-y., Shao, M.-a., Jaleel, C.A., Hong-mei, M., 2008. Higher plant antioxidants and redox signaling under environmental stresses. *Comptes Rendus Biologies* 331, 433-441.

Shirayama, H., Tohezo, Y., Taguchi, S., 2001. Photodegradation of chlorinated hydrocarbons in the presence and absence of dissolved oxygen in water. *Water Research* 35, 1941-1950.

Sinha, R.P., Häder, D.P., 2002. Life under solar UV radiation in aquatic organisms. *Advances in Space Research* 30, 1547-1556.

Sinha, R.P., Singh, S.P., Häder, D.-P., 2007. Database on mycosporines and mycosporine-like amino acids (MAAs) in fungi, cyanobacteria, macroalgae, phytoplankton and animals. *Journal of Photochemistry and Photobiology B: Biology* 89, 29-35.

Stahl, W., Sies, H., 2003. Antioxidant activity of carotenoids. *Molecular Aspects of Medicine* 24, 345-351.

Strid, Å., Chow, W., Anderson, J., 1994. UV-B damage and protection at the molecular level in plants. *Photosynthesis Research* 39, 475-489.

Teramura, A.H., 1983. Effects of ultraviolet-B radiation on the growth and yield of crop plants. *Physiologia Plantarum* 58, 415-427.

Tomaselli, L., Lamenti, G., Bosco, M., Tiano, P., 2000. Biodiversity of photosynthetic micro-organisms dwelling on stone monuments. *International Biodeterioration & Biodegradation* 46, 251-258.

Tretiach, M., Crisafulli, P., Imai, N., Kashiwadani, H., Hee Moon, K., Wada, H., Salvadori, O., 2007. Efficacy of a biocide tested on selected lichens and its effects on their substrata. *International Biodeterioration & Biodegradation* 59, 44-54.

Vardi, A., Berman-Frank, I., Rozenberg, T., Hadas, O., Kaplan, A., Levine, A., 1999. Programmed cell death of the dinoflagellate Peridinium gatunense is mediated by CO2 limitation and oxidative stress. *Current Biology* 9, 1061-1064.

Warscheid, T., Braams, J., 2000. Biodeterioration of stone: a review. *International Biodeterioration & Biodegradation* 46, 343-368.

Xiong, F., 2001. Evidence that UV-B tolerance of the photosynthetic apparatus in microalgae is related to the D1-turnover mediated repair cycle in vivo. *Journal of Plant Physiology* 158, 285-294.

Yakovleva, I.M., Titlyanov, E.A., 2001. Effect of high visible and UV irradiance on subtidal Chondrus crispus: stress, photoinhibition and protective mechanisms. *Aquatic Botany* 71, 47-61.

Zeeshan, M., Prasad, S.M., 2009. Differential response of growth, photosynthesis, antioxidant enzymes and lipid peroxidation to UV-B radiation in three cyanobacteria. *South African Journal of Botany* 75, 466-474.

In: Algae
Editors: Dagmar Krueger and Helga Meyer

ISBN: 978-1-62081-280-8

Chapter 2

Enhancement and Restoration of Growth of Durum Wheat (*Triticum durum*, Var. waha) on Saline Soil by Using *Azospirillum brasilense* NH and Marine Alga *Ulva lactuca*

Elhafid Nabti*[1], *Mohamed Sahnoune*[2], *Mostefa Ghoul*[3], *Doreen Fisher*[4], *Mike Rothballer*[4], *Michael Schmid*[4], and *Anton Hartmann*[4]

[1] University of Bejaïa, Laboratory of Applied Microbilogy, Targa Ouzemmour, Bejaïa, Algeria

[2] University of Bejaïa, Laboratory of Ecology and Environment, Targa Ouzemmour, Bejaïa, Algeria

[3] University of Setif, Laboratory of Bacterial Ecology, Setif, Algeria

[4] Helmholtz Zentrum Muenchen, German Research Center for Environmental Healt Department Microbe-Plant Interactions, Munich-Germany

Abstract

A new bacterial isolate *Azospirillum brasilense* NH strain isolated from salt-affected soil in the north of Algeria was applied to inoculate durum wheat (*Triticum durum*, var. waha). This strain showed already an important positive impact on durum wheat (stem height and spike length) in absence of salinity. The strain shows also a slight effect under saline conditions. Moreover, the combination of this stain with the marine alga *Ulva lactuca* illustrated the best results in absence or presence of high saltiness (160 and 200 mM NaCl). Proline and total sugar production (stress parameters) under saline conditions is inhibited by using *A. brasilense* NH, but the both components are strongly inhibited when *U. lactuca* was added. Also, rates of germination, stem height per week, height /width of the largest leaf, roots and shoot dry weight, optical density (OD) of chlorophyll a and b and total rate in chlorophyll, stem and roots dry weight, 1000-seed weight, total spike length, spike dry weight, seed number per spike and total weight of seeds per spike, were positively affected. It was shown that this new strain restores entirely the vegetable growth at 160 and 200 mM NaCl. This study was carried out using a reference strain *A.*

brasilense Sp7 in order to compare between the tow strains. It was shown that the reference strain Sp7 is totally inhibited at 200 mM NaCl.

Thus *A. brasilense* NH restored the vegetable growth of durum wheat and reduced the proline and total sugars under saline conditions in pot and field experiments. The results obtained proved the best effect of this new strain NH on durum wheat growth and inhibition of the osmotic stress. In additional to the restoration of durum wheat growth under saline conditions by *A. brasilense NH-U. lauca*, the algal extracts remains the good stimulator of plant growth in absence of stress.

The endophytic colonization of this strain was studied in presence or absence of salinity, in axenic conditions. Wheat roots were harvested and fixed with paraformaldehyde (4%) and hybridized using the FISH technique. Cofocal laser scanning microscopy (CSLM) was performed and the in situ localization of bacterial cells were obtained in the endophytic wheat roots. The best colonization was obtained at high salinity (200 mM NaCl). This result can justify the ability of this new strain to restore wheat growth at 200 mM NaCl. Thus, the results of our work display that the combination of *A. brasilense* NH – *U. lactuca* extracts could be a good solution to solve the problem of salinity which is the serious threat hindering a wheat culture in arid and saline soils.

Keywords: *Azospirillum brasilense*, Fertilization, wheat, saline stress, soil, CSLM

INTRODUCTION

Soil salinity constitutes a major obstacle for agriculture in arid and semi-arid regions. Actually, 20 % of the world's cultivated area and almost half of the world's irrigated soils are severely affected by the lack of water and increased salinity and are thus lost for agriculture. Many of these areas are confined to the tropical and subtropical Mediterranean areas [1, 2, 3].

To solve these problems, improvement of management practices by e.g. using suitable plant cultivars or osmotolerant, plant growth promoting rhizobacterial strains (PGPR) (e.g. *Pseudomonas* and *Azospirillum*) are being developed [4, 5, 6, 7]. In addition, supplements with natural substances (i.e. extracts of marine alga) were considered [8] allowing the restoration and improvement of salt-affected crop growth.

Cytoplasmatic membranes of organisms are permeable to water but not to other metabolites. Therefore, hyper- or hypo-osmotic shock exerted on cells causes a concomitant decrease or increase in the cytoplasmic volume leading to plasmolysis. The exposure of bacteria to high osmolarity conditions decreases water activity in their cytoplasm [9] and most of the cellular proteins and other biological macromolecules as well as essential functions are impaired [10]. Also, sudden plasmolysis inhibits various physiological processes, ranging from energy and nutrient uptake to inhibition of DNA replication and macromolecule biosynthesis [4, 11].

High salinity constitutes an environmental stress also for rhizospheric bacteria. In addition to general effects as outlined above, alteration of proteins involved in the initial attachment steps (adsorption and anchoring) of bacteria to plant roots in symbiotic interaction occur as well as inhibition of bacterial nodulation and nitrogen fixation activity, alteration of exopolysaccharide (EPS) and lipopolysaccharide (LPS) composition of the bacterial cell surface, impairment of molecular signal exchange between bacteria and their plant host due to the alteration of membrane glucan contents as well as inhibition of bacterial mobility and chemotaxis towards plant roots. High salinity decreases bacterial numbers colonizing root cells endophytically, because plants utilize proline, proline-betaine and glutamate under osmotic stress conditions and thus deprive rhizobacteria of these substances as energy and carbon sources as well as osmolytes which finally limits their growth. Depletion of potassium ions by plants reduces the ability of rhizobacteria to use potassium ions as primary osmoregulator [10, 12, 13].

Many studies reported that osmotic stress distorts plant´s physiology and cellular structure [4]. High salinity causes an inhibition and deterioration of several stages of plant life: germination, synthesis of phytohormones and other plant growth stimulating factors, photosynthesis, maturation of cell walls, plant morphology and elaboration, root and stem growth, ionic transport and nutrient uptake as well as general enzymatic activity [14].

Plant growth promoting rhizobacteria (PGPR) have a high potential for agriculture, because they can improve plant growth especially under limiting or stress conditions. For example, bacteria of the genus *Azospirillum* were successfully applied as inoculant in agriculture resulting in improved yields under severe conditions [15, 16, 17, 18, 19]. In *Azospirillum*, this growth promotion resulted from bacterial phytohormone synthesis, which promotes root development and the uptake of nutrients and water [20, 21, 22]. Water stress of plants was shown to be particularly diminished in wheat seedlings by inoculation with *Azospirillum brasilense* Sp245 [5]. Recently, an *Azospirillum brasilense* strain NH was isolated from salt-affected soil in northern Algeria [23]. We communicated the phylogenetic position and its most interesting halotolerant and halophilic properties, which exceeded the halotolerance of other reported halotolerant *A. brasilense* and other PGPR. While its growth was retarded at low salt concentrations, its optimal growth was at 200 mM NaCl in minimal medium without osmoprotective additives [23]. *A. brasilense* NH produces the auxin indole acetic acid (IAA) at osmotic stress conditions, when osmosensitive *A. brasilense* Sp7 showed only marginal IAA-production. Using glycine betaine and extracts of *Ulva lactuca*, its growth could be stimulated even at 300 to 500 mM NaCl [23].

Aqueous extracts of *Ulva lactuca* constitute an efficient source of osmoprotectants for microorganisms under saline stress. This alga contains high level of various betaines, amino acids, proteins and dimethylsulphoniopropionate (DMSP) [8, 24]. DMSP was demonstrated as potent osmoprotectant in several bacteria, such as *E. coli* [24], *Salmonella thyphimurium* [25], *Sinorhizobium meliloti* [26] and 40 % of marine prokaryotes [27]. DMSP is also an efficient osmoprotectant for agricultural bacteria [28, 29]. While the effect of salinity on plant growth was exhaustively studied [30, 31], only few data are yet available on the combined effects of osmotolerant *A. brasilense* strains and natural osmoprotectants (e.g., marine alga) on plant growth under saline stress in the field. The ability to promote growth of wheat under salinity stress by *A. brasilense* NH was reported recently only for two basic growth parameters [23].

In order to determine the impact of *A. brasilense* NH and/or of aqueous extracts of *U. lactuca* on the restoration and improvement of growth of durum wheat (*Triticum durum*, var. waha), a detailed study of plant performance was carried out in pots and field experiments under saline conditions. In addition, a laboratory monoxenic inoculation study with durum wheat and *A. brasilense* NH was carried out under saline and non saline conditions in order to reveal the mode of root colonization (potential for endophytic colonization) of *A. brasilense* NH using fluorescence *in situ* hybridization (FISH) analysis combined with confocal laser scanning microscopy (CLSM).

Materials and Methods

Plant and Bacteria Used

Durum wheat (*Triticum durum* var. waha), the predominantly used cultivar of wheat in the Bejaia area, Northern Algeria, was studied (INRA-Béjaïa-Algeria). The salt-tolerant bacterium *A. brasilense* NH, isolated from salt affected soil of a wheat field in the Bejaia area [23], was applied as inoculum. The salt-sensitive type strain *A. brasilense* Sp7 was used as control. *Burkholderia phytofirmans* PSJN was a gift from Dr. Angela Sessitsch (Seibersdorf, Austria).

Preparation of Algal Aqueous Extract

About 5 kg of green alga *Ulva lactuca* were harvested from the gulf of Bejaïa (Algeria) on March 2006. After washing with clean seawater, distilled water was added to make 500 g fresh weight of alga per liter. To release the osmoprotective substances, the alga aqueous suspension were autoclaved at 110°C / 30 min. The extracts had the following nutritive composition (g/100 g of dry matter): Glucids-36, lipids-0.8, protids-17 and minerals (K, P, N and Fe)-30 [32, 8].

Characteristics of the Experimental Site and Soil

The site of experimentation (Ihaddadhen, Bejaïa, Algeria) was located at 500 m of altitude and oriented north towards the sea. Chemical characteristics of the soil were as follows: organic matter (3.1 %), total N (0.48 %), mineral N (0.025 %), phosphate (0.1 %), potassium (0.13 %), Ca^{2+} (0.45 %) and Mg^{2+} (0.11 %), water holding capacity (25 %), pH 6.4, and 0.85 % salinity.

Experimental Design and Plant Culture

Three experiments were simultaneously conducted from March to June 2006 (i.e. from sowing to complete plant maturity on the 12th week after sowing) under a natural light/dark regime and temperatures ranging from 25-32°C during the experimentation period. Pot experiment 1 and pot experiment 2 were performed at an initial NaCl concentration of 100 and 150 mM, respectively. In the third week after sowing the salinity was increased to a final concentration of 160 and 200 mM NaCl respectively. Each pot experiment consisted of 10 treatments (T1-T10) with 50 plants each (with finally 1 plant per pot); pot experiment 1 and pot experiment 2 differed in six treatments T3-T8, where NaCl was involved (Table 1). 1.5 l pots were filled with autoclaved soil (110°C /30 min) taken from the experimental site described above. Surface disinfection of seeds was performed by immersion in 1 % of NaOCl solution followed by thorough washing in distilled water. Five surface-disinfected seeds were sown per pot; the 5 seeds were put together 3 cm deep into the soil at the centre of the pot.

Algal aqueous extracts (100 ml, 500 g/l) and/or 10 ml of bacterial suspension (10^6 CFU/ml) were added to the appropriate seeds immediately after sowing. No fertilizer was applied throughout the entire experiment. After emergence of the first leaf, the plant density was reduced to one plantlet per pot. The positions of the 800 pots of the two pot experiments were completely randomized to exclude any location effect on plant growth. The total randomization was carried out using STAT-ITCF package [33]. To avoid cross-contamination, the pots were placed 30 cm apart and the soil surface was covered with a 3 cm thick layer of sterile vermiculite. Pots were wrapped in black plastic in order to mitigate the effects of light. All potted plants were isolated from precipitation water using a greenhouse.

Table 1. Composition of the 10 plots used in pots and field experiments

N°	Pots		Field
	Experiment (1)	Experiment (2)	
1	Wheat	Wheat	Wheat
2	Wheat + U	Wheat + U	Wheat + U
3	Wheat + 160 *	Wheat + 200 *	Wheat + 200 *
4	Wheat +160+U	Wheat 200+ U	Wheat 200+ U
5	Wheat+160 +NH	Wheat+200+NH	Wheat+200+NH
6	Wheat +160+Sp7	Wheat +200+Sp7	Wheat +200+Sp7
7	wheat+160+U+NH	Wheat+200+U+NH	Wheat+200+U+NH
8	wheat+160+U+Sp7	Wheat+200U+Sp7	Wheat+200U+Sp7
9	Wheat +Sp7	Wheat +Sp7	Wheat +Sp7
10	Wheat +NH	Wheat +NH	Wheat +NH

U: algal aqueous extract of (*U. lactuca*). NH: *A. brasilense NH*. Sp7: *A. brasilense Sp7*. * : 160 and 200 mM NaCl.

The field experiment was carried out on an area of 225 m^2. The cultivation procedure was as in the pot experiments, except that the seeds were placed 5 cm deep holes in the soil surface. For each treatment 30 plants were sown in a single line with 80 cm between plants and 75 cm between lines, soil surface not covered with vermiculite. Altogether, the field experiment consisted of 300 not randomized plants.

Plant Irrigation

For all three experiments, each pot (pot experiment 1 and pot experiment 2) or a hole in the ground (field experiment) was irrigated once a week with 500 ml of non-saline water from a nearby well. In the first three weeks, the pots or holes in the ground of the treatments where salt was involved were irrigated with 500 ml of 100 mM NaCl (pot experiment 1) or 150 mM NaCl (pot experiment 2 and field experiment) instead of water. The initial NaCl concentrations were increased at the third week after sowing for all the three experiments from 100 and 150 mM to 160 and 200 mM NaCl respectively. The salinity in the treatments to which salt was not added was about 1.5 mM NaCl. The effective salt concentrations were measured using a conductivity meter. The cultures were so entertained until complete plant maturity (12th week after sowing).

Measured Plant Parameters

The respectively 50 and 30 sets of 5 seeds per treatment described above for the pot and field experiments were used to determine the germination rates, expressed as confidence intervals (P=0.05) of mean numbers of germinations (M±2SE) observed over the nine first days after sowing. In addition, 17 other parameters were measured on grown plants (9-12th week after sowing).

These parameters (numbered by the left side exponents) listed as follows: ^{1}stem height (cm), ^{2}length (cm) and ^{3}width (mm) of the largest leaf, ^{4}stem dry weight (g), ^{5}roots dry weight (g), ^{6}total spike length (beard hairs included) (cm), ^{7}beard hairs length (cm), ^{8}spike length (beard hairs not included) (cm), ^{9}spike dry weight (g), ^{10}seeds number per spike, ^{11}total seed weight per spike, ^{12}1000-seeds weight (g); and, on mature green leaves, ^{13}chlorophyll a, ^{14}chlorophyll b and ^{15}total chlorophyll rates/fresh matter (/FM) (mg/g); ^{16}proline content/FM (µg/g [34] and ^{17}total sugars rate/FM (µg/g). The 1-12th parameters were measured of all experimental plants, but chlorophylls, proline and sugar contents (13-17th parameters) were determined only for 10 plants per treatment.

Statistical Analysis

The values of the measured parameters are expressed as means ± standard deviation (M±SD) or means ± 1.96(2) standard error (M±2SE).

For statistical comparison of these mean values, one-way anova and Fisher's least significant difference tests were applied (P=0.05). The 10 treatments within each experiment, and the 3 experiments within each treatment were compared for all the measured parameters separately (monofactorial comparison).

For multifactorial comparison, principal component analysis (PCA) was used to display the correlations between the various plant parameters and the affinities between the different treatments within and between experiments. For this, a PCA was performed for each experiment separately and then another PCA for all the 3 experiments taken together. The position of the treatments in the PCA projections were also tested using Anova and Fisher's LSD tests (P=0.05).

The programs used to achieve all the statistical analyses were the Statistica 7.0 package [33] PCA analyses included 10 plants per treatment and all 17 measured parameters.

In Situ Localization of *A. brasilense* NH in Wheat Roots with FISH Analysis and Confocal Laser Scanning Microscopy (CLSM) in a Monoxenic System

Wheat seed sterilization: 30 wheat seeds were shaken vigorously for 2 min in ethanol (70 %, v/v), and then treated for 15 min with NaOCl (2 %) solution, washed several times with sterile distilled water and finally treated for 1 h in a solution with a mixture of antibiotics (penicillin 600 mg/ml, streptomycin 250 mg/ml, cycloheximide 100 mg/ml).

Finally, seeds were germinated on nutrient broth plates at 30°C in the dark to display possible contaminations [35]. Inoculum preparation: Liquid NB medium was inoculated with *A. brasilense* NH and incubated overnight with shaking at 30°C. The culture was centrifuged at 6000 × g for 5 min. and the pellet was washed twice with 20 ml sterile PBS (phosphate buffered saline).

The bacterial suspension used for the experiment was 10^8 cells/ml ≈ OD 0.7 at 436 nm. Monoxenic plant growth system: Test tubes (200 ml volume) were filled up to one third of their volume with quartz sand and 10 ml of a plant nutrient solution Length Ashton (Sigma) (1 g/100 ml) was added and autoclaved at 121°C for 20 min. NaCl was added to the appropriate tubes at 200 mM concentration.

Wheat seedlings, germinated under sterile conditions, were washed with sterile distilled water and incubated with the bacterial suspension for 1h at room temperature. In each tube, one seed was planted in a depth of 1 cm. The glass tubes were exposed to day light for growth for 30 days at room temperature.

Preparation of root samples: After 30 days, seedlings were carefully removed from the tubes and washed by shaking in sterile PBS. Root pieces 15 to 30 mm in length from all parts of the root system were transferred into the fixation buffer (paraformaldehyde 4 % in PBS) and incubated at room temperature for 2 h.

Samples were washed and dehydrated by incubating for 5 min each in an increasing ethanol series at 50, 80 and 96 %. Each root piece was immobilized on gelatin-coated slides with a droplet of agarose (0.25 %) on both ends.

In situ hybridization and microscopy: The following oligonucleotide probes were used: Eub-338-I [36], II and III [37] as equimolar mixture labeled with the fluorescent dye Fluorescein (FITC), complementary to a region of the 16S rRNA specific for the domain Bacteria, and Abras-1420-Cy3, specific for the species *A. brasilense* [38]. Coated slides already prepared for hybridization were used.

Each spot received 15 μl of hybridization buffer [(NaCl (0.9 M), 20 mM, Tris-HCl [pH 7.2], SDS (0.01 %), EDTA (5 mM, deionized formamide (20 %)]. 2 μl of the probes were added. Incubation was performed at 46°C for 2 h. The slides were transferred to the washing buffer (20 mM Tris; 0.01 % SDS; 5 mM EDTA; 0.09 M NaCl) and then incubated at 48°C for 20 min. Finally, the slides were rinsed with bidistilled water and dried in an air stream.

Fluorescing cells on or in roots were detected by a confocal laser scanning microscope (LSM-510 MEta, Zeiss, Oberkochen, Germany).

An Argon Ion Laser supplied the wavelength of 488nm for excitation of FITC, and two Helium Neon Lasers provided excitation wavelengths of 543nm and 633nm for Cy3 and Cy5 excitation, respectively. A water immersion lens C-Apochromat 63x / 1.2 W (Zeiss, Oberkochen, Germany) was used for observation.

Detection of ACC-Deaminase Activity

The 1-aminocyclopropane-1-carboxylate (ACC) deaminase activity of bacteria was detected [39] using *Burkholderia phytofirmans* PSJN as positive reference culture.

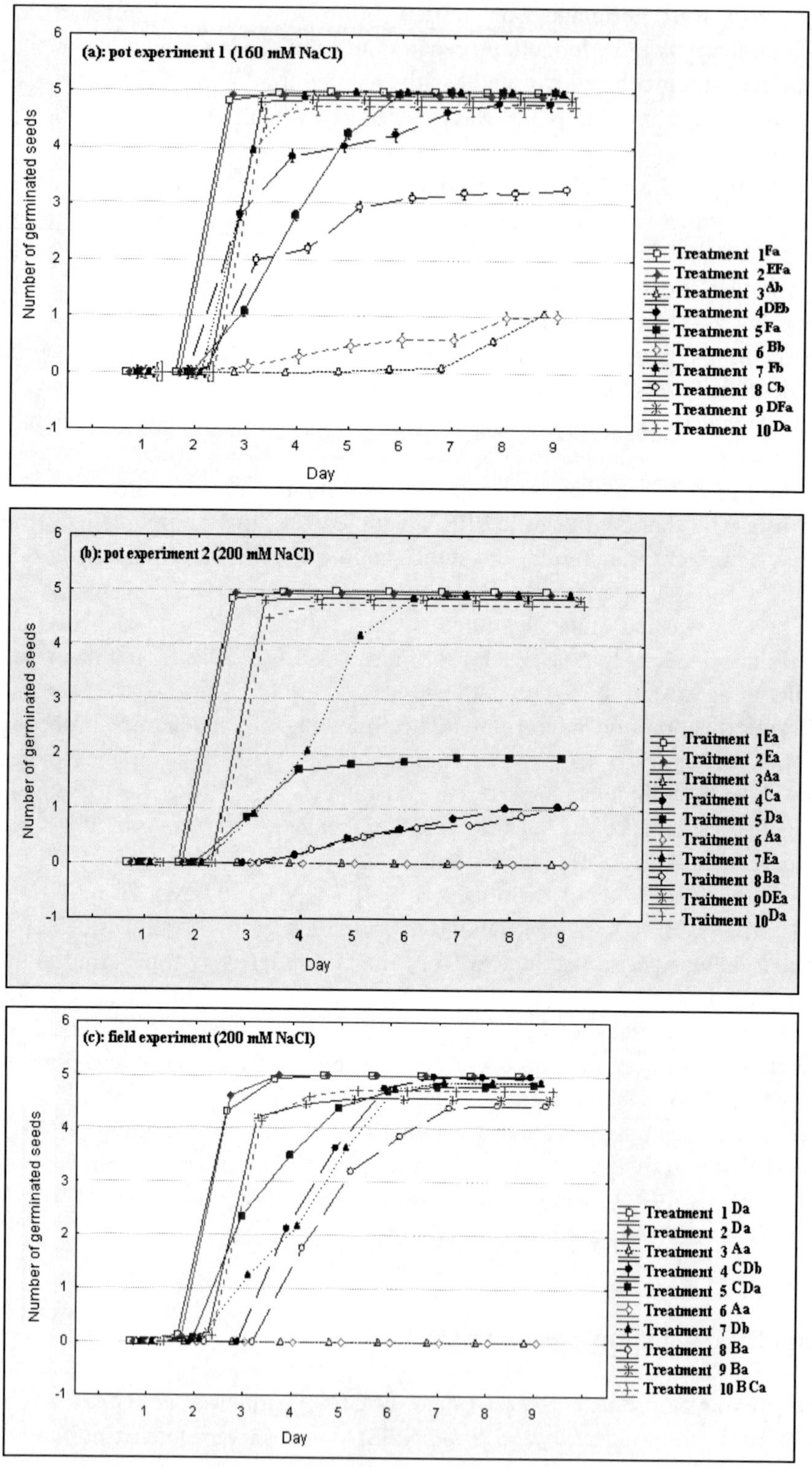

Figure 1. Germination rate of wheat seeds during nine days.

RESULTS

Effects of Various Treatments on Germination under Saline and Non-Saline Conditions

The germination rates of wheat seeds were greatly influenced by salt stress. They were either inhibited by salt or restored partially to completely following the experimental treatments (Figure 1). In the presence of 160 mM NaCl (treatment 3) the germination rate was almost completely inhibited. The inoculation with *A. brasilense* NH together with the application of algal extracts (treatment 7) resulted in the same high numbers of germinated seeds as compared to the control without salt stress (treatment 1). The addition of algal extracts or *A. brasilense* NH alone (treatments 4 and 5) gave only partial recoveries. The inoculation with *A. brasilense* Sp7 reduced the NaCl effect only very slightly (treatment 6, Figure 1 a).

When 200 mM NaCl was applied in pot and field experiments, the germination rate was completely inhibited and could not be restored by inoculation with *A. brasilense* Sp7 (treatments 3 and 6, Figure 1 b and c). The inoculation with *A. brasilense* NH (treatment 5) restored the germination only partly and supported the germination of about a third of the seeds. The combination of inoculation with the strain NH and treatment with *U. lactuca* extract promoted the germination of all seeds (treatment 7) although with a delay of about 4 days as compared with the control treatments without salt stress (Figure 1 c). On the other hand, wheat seeds inoculated with *A. brasilense* Sp7 were still greatly affected by salt stress and Sp7 did not reduce the stress effect on germination at 200 mM NaCl (treatment 4 and 8, Figure 1 b and c).

Effects of the Various Treatments on Growth and Yield Parameters under Saline and Non-Saline Conditions

Under non saline conditions (treatments 1, 2, 9, and 10), the changes observed are small but significant in some cases. Stem height ([1]SH) was increased slightly in the presence of the algal extracts (treatment 2) and in the presence of the two strains of *Azospirillum brasilense* Sp7 (treatment 9) or NH (treatment 10), as compared to the control without bacterial inoculation (treatment 1) (Table 3). The stimulatory effect of both *A. brasilense* Sp7 and NH inoculation (treatment 9 and 10) on leaf growth (Leaf length ([2]LL) and Leaf width ([3]LW)) was more significant than the effect of algal extracts (treatment 2). Stem dry weight ([4]ADW) was more improved by the addition of algal extracts then inoculation with the strains Sp7 or NH, respectively; the control plants always presented the lowest [4]ADW value. Root dry weight ([5]RDW) was more stimulated by the inoculation with both Sp7 and NH than by the addition of the algal extracts. Total spike length ([6]TSL) was increased in plants treated with algal extracts (treatment 2) as well as in the inoculated plants (treatment 9 and 10) as compared to the control. The spike length (without hairs) ([8]SL) and spike hair length alone ([7]HL) showed similar effects as in the case of [6]TSL. The spike dry weight ([9]SDW) was only slightly increased in the presence of algal extracts and the bacterial inocula. In the pot experiments, inoculation with strain Sp7 promoted [9]SDW slightly more than with the NH strain. The number of seeds per spike ([10]SNS) was only affected in the presence of algal extracts, while the inoculation with Sp7 or NH did not result in any difference from the control. In contrast, total weight of seeds per spike ([11]SWS) in the pot experiments was increased after the addition of algal extracts and the inoculation with *A. brasilense* Sp7, while *A. brasilense* NH was less efficient in this case. In the pot experiments, the most important yield parameter (1000-seeds weight, [12]1000SW) was also more stimulated in wheat plants inoculated with *A. brasilense* Sp7, as compared to inoculation with the NH strain or in the non inoculated control plants (with and without algal extracts). The chlorophyll a ([13]CHA), b ([14]CHB) as well as the total contents of chlorophyll ([15]CHT) were slightly higher when the algal extracts were added as was the case after inoculation with Sp7 or NH (Table 3). While the sugar content ([17]TS) was significantly reduced in inoculated plants (especially in the pot experiments) the proline content ([16]PRO) was only slightly affected by inoculation or addition of algal extracts (treatments 1, 2, 9 and 10).

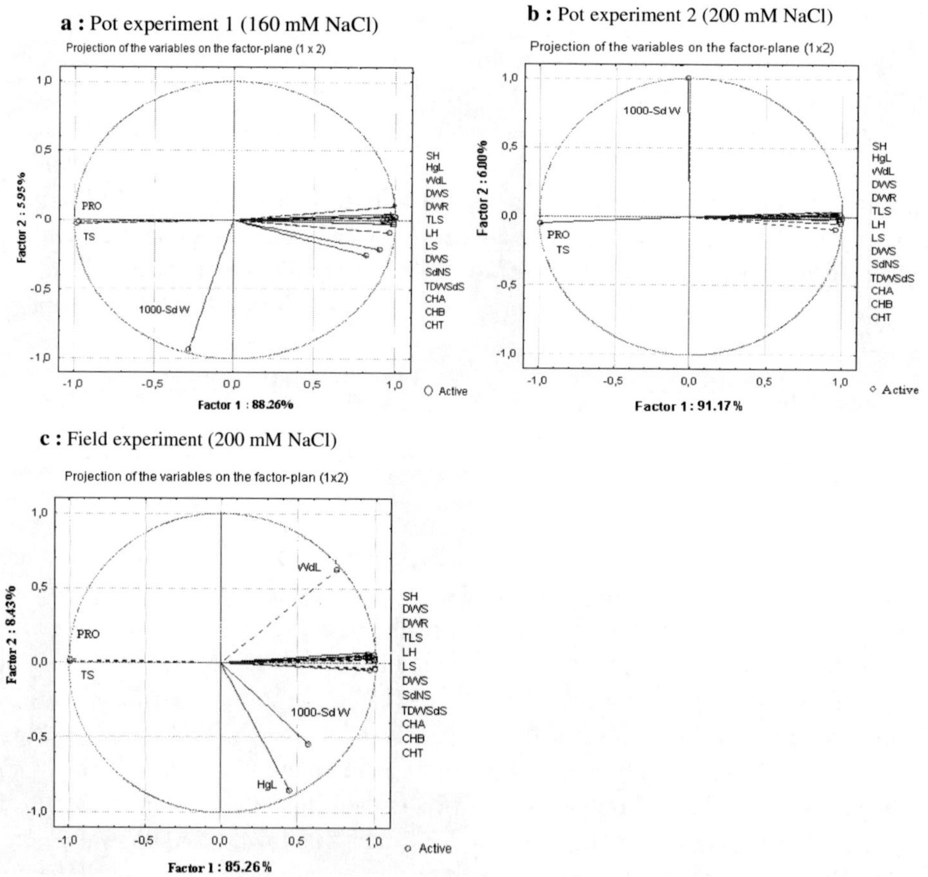

Figure 2. Correlation circle for Principal Component Analysis (PCA) of (a) pot experiment 1 (160 mM NaCl), (b) pot experiment 2 (200 mM NaCl) and (c) field experiment (200 mM NaCl). For the code variables see Table 2.

Wheat growth was inhibited partially or completely in the presence of 160 and 200 mM NaCl in the pot and field experiments (treatment 3). The inoculation with *A. brasilense* Sp7 alone or together with algal extracts could not recover salt-affected plant growth (treatments 6 and 8) (Table 3). In contrast, the addition of *A. brasilense* NH alone and even more together with algal extracts (treatments 5 and 7) recovered wheat growth partially or completely (Table 3). As can be seen from the comparison of the treatments 3 and 4 (salt inhibition without and with addition of *U. lactuca* extracts), algal extracts can recover to some extent salt-affected plant growth. The combination of inoculation with *A. brasilense* NH and addition of algal extracts under salt stress (160 or 200 mM) was more efficient than the separate application of *A. brasilense* NH or *U. lactuca* extracts (treatments 4, 5 and 7). Most plant parameters tested, like stem height, leaf length and width, stem and root dry weight, length and dry weight of spikes, seed numbers per spike and total weight of seeds per spike were all responding very similarly. In addition, the chlorophyll content of leaves presented the same response pattern. However, the 1000-seeds weight revealed, that if the plants could grow (they failed to grow at all at 200 mM NaCl with and without inoculation with Sp7; treatments 3 and 6) they reached about the same weight, the inoculation with Sp7 giving the highest values (Table 3).

The contents of proline and total sugars were found greatly elevated in the presence of salt stress. These contents increased from treatment 7 to treatment 3 and 6 (7<5<8<4<3 and 6) and were proportional to the stress intensity. The results of the durum wheat growth parameters under saline stress (160 or 200 mM NaCl) in pots and field experiments demonstrated that aqueous extracts of *U. lactuca* provide powerful osmoprotectants for the growth of durum wheat (*T. durum* var. waha), but their effect was less marked than the inoculation with *A. brasilense* NH.

Correlation analyses clearly indicate relationships between the parameters measured in the treatments 1-10 in the pot and field experiments. The principal component analysis (PCA) and their correlation were shown in Figure 2 a-d and Table 3.

Parameters like stem height, leaf height and width, stem and roots dry weights, spike length and/or hairs, spike dry weight, number of seeds per spike and total weight of seeds per spike were clustering together, while the contents of proline and total sugar are located at the opposite axis (Figure 2). Therefore, these two parameters were inversely proportional to the plant growth. In fact, when the plant was grown under osmotic stress, proline and total sugars contents increased, while other parameters were inhibited or decreased (Figure 2 a-d). In contrast, 1000-seeds weight (1000 SW) was not correlated with the before-mentioned parameters, reflecting a very different regulatory mechanism controlling seed weight. The distribution of parameters of various treatments (1-10) revealed clearly that the measured values of the treatments 1, 2, 7, 9, and 10 were more significantly clustering than those of treatments 3, 4, 5, 6, and 8 (Figure 2 a-d).

There was a strong correlation between all the parameters measured (including leaf height and leaf width) except the 1000-seeds weight (Figure 2 a-d). 1000-seeds weight was superior in the pots containing high NaCl level; as the plant would preserve its seed, 1000-seeds weight is proportional to the osmotic stress (Figure 2 a-d). However, in pot experiment 2 (200 mM NaCl), 1000-seeds weight is more correlated with Factor 1 and less correlated with Factor 2 (Figure 2 b) than in pot experiment 1 (160 mM NaCl) and the field experiment (200 mM NaCl) (Figure 2 a and c). This suggests that salinity stress on leaf parameters was stronger in pot experiments than in the field experiment.

Principal component analysis revealed, that the results obtained in the field experiment at 200 mM NaCl cluster in two groups: the first group contains the treatments 1, 2, 9 and 10, where no salt stress was applied, and the second group (treatments 4, 5, 7 and 8) with salt stress (Figure 3 c). The treatments 3 (NaCl) and 6 (NaCl + Sp7) are missing, because growth was completely abolished. Moreover, the similarity values obtained under various treatments in the field experiment were higher than those obtained in the pot experiments at 160 and 200 mM NaCl, respectively (Figure 3 a, b). The treatment with *U. lactuca* extracts under 200 mM NaCl stress in the pot experiment 2 is especially separated from the other value groups (Figure 3 b).

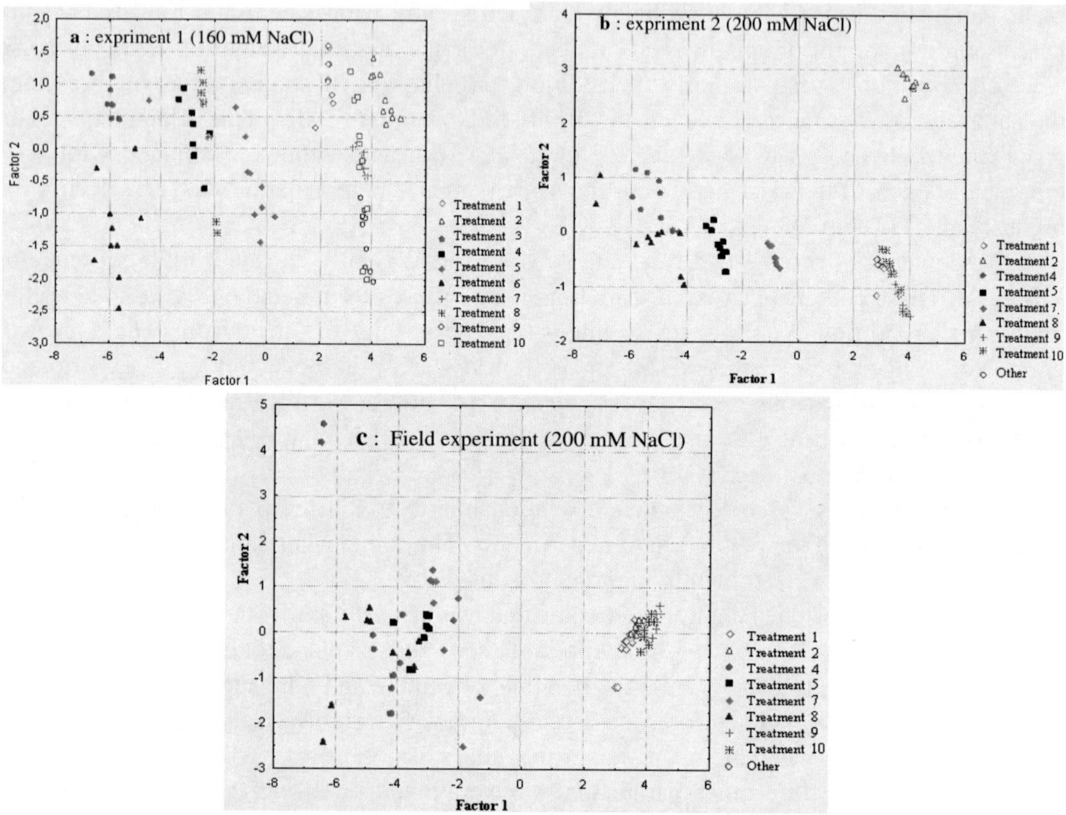

Figure 3. Principal Component Analysis (PCA) for pot experiment 1, 2 and field experiment (160, 200 mM NaCl). Treatments 3 and 6 were not included in figures b and c because plants have not grown under these conditions. For the treatments codes see Table 3.

In Situ Localization of *A. brasilense* NH in Wheat Roots Using the FISH Analysis and Confocal Laser Scanning Microscopy (CLSM)

In the monoxenic quartz sand culture conditions the wheat seedlings could only grow at 200 mM NaCl in the presence of *A. brasilense* NH. The *in situ* localization of *A. brasilense* NH using two oligonucleotide probes (EUB-338-I, II, III-FLUOS and Abras-1420-Cy3) concomitantly resulted in a clear localization and identification of single cells and small cell clusters on the root surface but also frequently within the roots (Figure 3a, b, c, d). The predominant endophytic localization was the apoplastic space between the root epidermal cells.

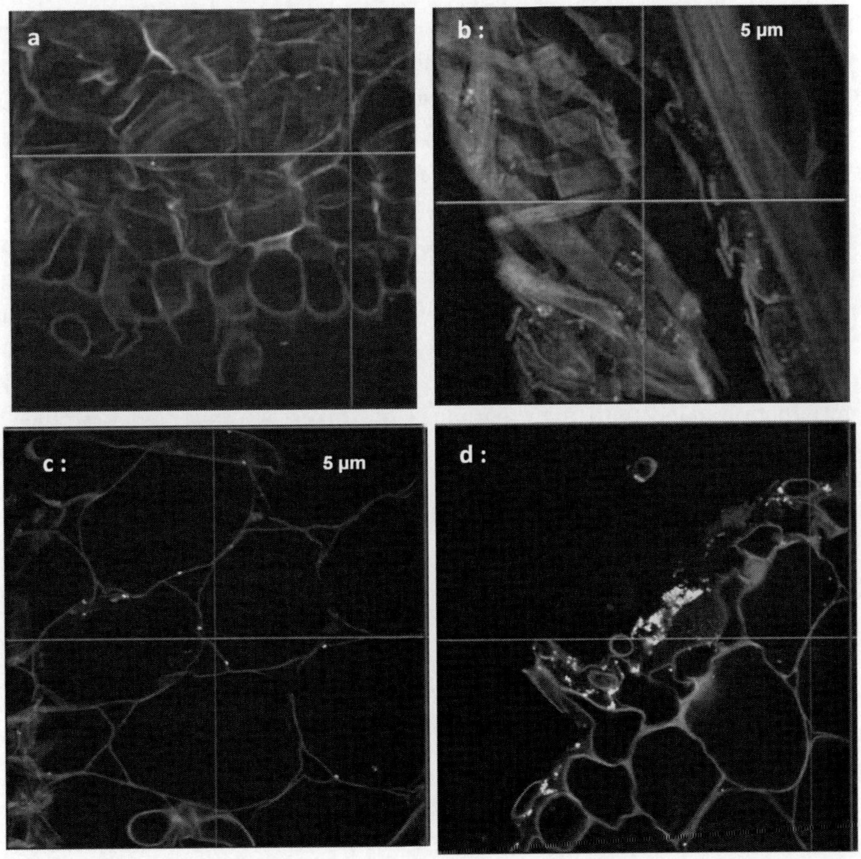

Figure 4. Confocal laser scanning microscopic images of wheat roots inoculated with *A. brasilense* NH in axenic conditions after 4 weeks of growth under saline conditions (200 mM NaCl): The roots were fixed in 4 % PFA, FISH-analysis was performed using the probes EUB- 338-I, II, III-FLUOS and Abras-1420-Cy3. *A. brasilense* NH cells are stained in yellow (combination of both of the fluorescent signals [Cy3=red, FLUOS=green]). A: Endophytic colonization. Orthogonal optical sections of a three dimensional confocal image created from a *z*-stack of *xy*-scans. The top view, framed in blue, shows one picture from the center of this z-stack. The red and green lines represent vertical or horizontal optical cuts through the stack, which result in the side view images framed in red and green, respectively. B: *In situ* detection of *A. brasilense* NH colonizing the root hair zone. C, D: Radial slice from inoculated wheat roots. C: Colonization of intercellular spaces. D: Surface colonization

ACC-Deaminase Activity of *A. brasilense* NH

Using the growth assay on minimal medium plates with aminocyclopropane-1-carboxylate (ACC) as sole nitrogen source, *A. brasilense* NH failed to grow in contrast to *B. phytofirmans* PSJN and other ACC-positive bacteria.

Discussion

The data presented demonstrate the exceptional ability of *A. brasilense* NH alone or in combination with aqueous extract of the alga *Ulva lactuca* to restore and improve growth of durum wheat (*Triticum durum* var. waha) under saline stress conditions. Under non-saline conditions, the aqueous extract of the marine alga *U. lactuca* was also able to support wheat growth to some extent (Table 3). This can be explained by the nutritive value of the extract (see Material and Methods) and provides evidence that the composition and application rate of the algal extracts could improve the fertility of poor soils by providing it with P, N and other essential growth elements [40, 41]. In addition, both strains of *A. brasilense* (Sp7 and NH) could slightly stimulate wheat growth from the germination to the harvest stage also under non-saline conditions (Table 2). The promotion of wheat growth by *A. brasilense* is probably due to two mechanisms: the production of phytohormones like indole acetic acid [42] and the stimulation of ß-glucuronidase in wheat roots [43, 44, 45]). Moreover, *Azospirillum* is able to improve mineral uptake and to increase the nitrogen assimilation rate from the germination stage until the emergence of spikes [46, 47]. On the other hand, the PGPR-effect of *A. brasilense* Sp7 was noticed only in the absence of salt stress because *A. brasilense* Sp7 is NaCl-sensitive. It had been demonstrated, that salt stress altered severely the attachment of *A. brasilense* Cd, a very close relative to strain Sp7, to wheat roots by inhibiting both steps of adsorption and the anchoring, which was correlated to the impairment of the exopolysaccharide, glucane and lipopolysaccharide contents [12]. In addition, phytohormone biosynthesis and nitrogenase activity of *A. brasilense* Sp7 were inhibited under osmotic stress [48]. In contrast, *A. brasilense* NH reached its optimum of growth and indole acetic acid production at 200 mM NaCl [23].

Inoculation with *Azospirillum brasilense* NH improved germination at saline conditions greatly (Figure 1) while application of algal extract had only a slight effect on germination rate. Probably algal extracts provided osmolytes to the seed and thus retained water activity in the seeds [41]. Therefore, the combination of *A. brasilense* NH and algal extracts resulted in greatly improved germination compared to the control (treatments 1 and 7). The synergistic effect of these two treatments apparently improved the germination conditions to an extent as in the absence of salt stress (Table 2). The results of partial inhibition of the shoot and root development of durum wheat at 160 mM NaCl or total inhibition at 200 mM NaCl agreed with those obtained [49, 50]. Osmotic stress reduced stem height, root length and their dry weight. When environmental NaCl concentrations increase, sodium uptake into roots raises as well, while N, P, K and Mg-uptake decreases and intracellular ionic equilibrium is disturbed [51, 52]. The principal components analysis (PCA) of the growth data of wheat in the pot experiments with 200 mM NaCl revealed that the leaf length and width were not reduced to the same level as other parameters (Figure 2 b). When wheat plants are cultivated under saline or water stress, the leaf responded initially by changes in the volume and the number of stomata in order to maintain sufficient water to restore the photosynthetic activity. The leaf should thus keep its surface intact to ensure good plant growth and development [53]. The exposure of wheat seeds (*Triticum aestivum* L.) or barley (*Hordeum vulgare* L.) to 100 mM NaCl reduced the growth of their roots (length and dry weight) by 20 %, whereas leaves remained intact [54]. On the other hand, stem growth is very sensitive to osmotic stress [55]. Similar results concerning stem, leaf and root growth were obtained [55, 56]. The 1000-seeds weight did not correlate with other growth parameters (PCA) (Figure 2 a-d). Wheat seedlings affected by saline stress try to preserve their reserves (seed), whereas other parameters are altered [57, 58, 59]. When wheat plants were grown at high salinity conditions, stem and root growth as well as the number of spikes and seeds were affected, but the 1000-seeds weight remained more or less unchanged [60, 56, 61].

Table 2. Germination rate of the wheat seeds (5 seeds) for the 3 experiments

N°	Treatment	Day 3			Day 4		
		160 mM NaCl	200 mM NaCl	Field (200) mM NaCl	160 mM NaCl	200 mM NaCl	Field (200) mM NaCl
1	Control	4.82±0.59^{Ga}	4.82±0.59^{Da}	4.82±0.59^{DEa}	4.96±0.19^{Ga}	4.96±0.19^{Ea}	4.96±0.19^{Fa}
2	U	4.92±0.27^{Ga}	4.92±0.27^{Da}	4.92±0.27^{Ea}	4.92±0.27^{Ga}	4.92±0.27^{Ea}	4.92±0.27^{Fa}
3	NaCl	0.00±0.00^{Aa}	0.00±0.00^{Aa}	0.00±0.00^{Aa}	0.00±0.00^{Aa}	0.00±0.00^{Aa}	0.00±0.00^{Aa}
4	NaCl+U	2.80±0.53^{Db}	0.00±0.00^{Aa}	0.00±0.00^{Aa}	3.84±0.55^{Ec}	0.14±0.35^{Aa}	0.14±0.35^{Cb}
5	NaCl+NH	1.08±0.44^{Bb}	0.84±0.37^{Ba}	0.84±0.37^{Cc}	2.78±0.46^{Db}	1.70±0.50^{Ba}	1.70±0.50^{Da}
6	NaCl+Sp7	0.10±0.30^{Ab}	0.00±0.00^{Aa}	0.00±0.00^{Aa}	0.28±0.45^{Bc}	0.00±0.00^{Aa}	0.00±0.00^{Ab}
7	NaCl+NH+U	3.96±0.35^{Eb}	0.90±0.36^{Ba}	0.90±0.36^{Bc}	4.92±0.27^{Gb}	2.06±0.37^{Ca}	2.06±0.37^{Cc}
8	NaCl+Sp7+U	2.00±0.40^{Cb}	0.00±0.00^{Aa}	0.00±0.00^{Aa}	2.20±0.40^{Cb}	0.14±0.35^{Aa}	0.14±0.35^{Ba}
9	Sp7	4.80±0.67^{Ga}	4.80±0.67^{Da}	4.80±0.67^{DEa}	4.84±0.65^{FGa}	4.84±0.65^{DEa}	4.84±0.65^{Ea}
10	NH	4.50±0.88^{Fb}	4.50±0.89^{Cb}	4.50±0.88^{Daa}	4.74±0.66^{Fb}	4.74±0.66^{Db}	4.74±0.66^{Ea}
N°	Treatment	Day 5			Day 6		
		160 mM NaCl	200 mM NaCl	Field (200) mM NaCl	160 mM NaCl	200 mM NaCl	Field (200) mM NaCl
1	Control	4.98±0.14^{Ga}	4.98±0.14^{Fa}	4.98±0.14^{Ea}	4.98±0.14^{Fa}	4.98±0.14^{Ea}	4.98±0.14^{Ea}
2	U	4.92±0.27^{Ga}	4.92±0.27^{Fa}	4.92±0.27^{Ea}	4.92±0.27^{Fa}	4.92±0.27^{Ea}	4.92±0.27^{Ea}
3	NaCl	0.02±0.14^{Aa}	0.00±0.00^{Aa}	0.00±0.00^{Aa}	0.06±0.24^{Aa}	0.00±0.00^{Aa}	0.00±0.00^{Aa}
4	NaCl+U	4.02±0.38^{Dc}	0.42±0.49^{Ba}	0.28±0.45^{Cb}	4.22±0.42^{Db}	0.42±0.49^{Ba}	0.42±0.49^{CDEc}
5	NaCl+NH	4.24±0.72^{Ec}	1.86±0.35^{Cb}	1.80±0.40^{DEa}	4.94±0.24^{Fb}	1.86±0.35^{Cab}	1.86±0.35^{DEa}
6	NaCl+Sp7	0.48±0.50^{Bb}	0.00±0.00^{Aa}	0.00±0.00^{Ac}	0.58±0.49^{Bb}	0.00±0.00^{Aa}	0.00±0.00^{Ac}
7	NaCl+NH+U	4.98±0.14^{Gb}	4.86±0.35^{Da}	4.16±0.55^{Cc}	4.98±0.14^{Fb}	4.86±0.35^{DEa}	4.86±0.35^{CDEb}
8	NaCl+Sp7+U	2.94±0.24^{Cb}	0.56±0.50^{Ba}	0.32±0.47^{Ba}	3.10±0.30^{Cb}	0.56±0.50^{Ba}	0.56±0.50^{Ba}
9	Sp7	4.84±0.65^{FGa}	4.84±0.65^{EFa}	4.84±0.65^{Da}	4.84±0.65^{EFa}	4.84±0.65^{DEa}	4.84±0.65^{Ca}
10	NH	4.74±0.66^{Fa}	4.74±0.66^{Ea}	4.74±0.66^{DEa}	4.74±0.66^{Ea}	4.74±0.66^{Da}	4.74±0.66C^{Da}
N°	Treatment	Day 7			Day 8		
		160 mM NaCl	200 mM NaCl	Field (200) mM NaCl	160 mM NaCl	200 mM NaCl	Field (200) mM NaCl
1	Control	4.98±0.14^{Fa}	4.98±0.14^{Ea}	4.98±0.14^{Ea}	4.98±0.14^{Fa}	4.98±0.14^{Ea}	4.98±0.14^{Da}
2	U	4.92±0.27^{Fa}	4.92±0.27^{Ea}	4.92±0.27^{Ea}	4.92±0.27^{EFa}	4.92±0.27^{Ea}	4.92±0.27^{Da}
3	NaCl	0.08±0.27^{Ab}	0.00±0.00^{Aa}	0.00±0.00^{Aa}	0.10±0.30^{Ab}	0.00±0.00^{Aa}	0.00±0.00^{Aa}
4	NaCl+U	4.62±0.49^{Db}	0.54±0.50^{Ba}	0.54±0.50^{Ec}	4.78±0.42^{DEb}	0.56±0.50^{Ba}	0.56±0.50^{Db}
5	NaCl+NH	4.98±0.14^{Fa}	1.92±0.27^{Ca}	1.92±0.27^{DEa}	4.98±0.14^{Fa}	1.92±0.27^{Ca}	1.92±0.27^{CDa}
6	NaCl+Sp7	0.60±0.49^{Bb}	0.00±0.00^{Aa}	0.00±0.00^{Ac}	0.82±0.39^{Bb}	0.00±0.00^{Aa}	0.00±0.00^{Ac}
7	NaCl+NH+U	4.98±0.14^{Fb}	4.92±0.27^{Ea}	4.92±0.27^{Eb}	4.98±0.14^{Fb}	4.92±0.27^{Ea}	4.92±0.27^{Db}
8	NaCl+Sp7+U	3.18±0.39^{Cb}	0.60±0.49^{Ba}	0.60±0.49^{Ba}	3.18±0.39^{Cb}	0.62±0.49^{Ba}	0.62±0.49^{Ba}
9	Sp7	4.84±0.65^{EFa}	4.84±0.65^{DEa}	4.84±0.65^{BCa}	4.84±0.65^{DEFa}	4.84±0.65^{DEa}	4.84±0.65^{Ba}
10	NH	4.74±0.66^{DEa}	4.74±0.66^{Da}	4.74±0.66C^{Da}	4.74±0.66^{Da}	4.74±0.66^{Da}	4.74±0.66^{BCa}
N°	Treatment	Day 9					
		160 mM NaCl	200 mM NaCl	Field (200) mM NaCl			
1	Control	4.98±0.14^{Fa}	4.98±0.14^{Ea}	4.98±0.14^{Da}			
2	U	4.92±0.27^{EFa}	4.92±0.27^{Ea}	4.92±0.27^{Da}			
3	NaCl	0.12±0.33^{Ab}	0.00±0.00^{Aa}	0.00±0.00^{Aa}			
4	NaCl+U	4.78±0.42^{DEb}	0.58±0.49^{Ba}	0.58±0.49^{CDb}			
5	NaCl+NH	4.98±0.14^{Fa}	1.92±0.27^{Ca}	1.92±0.27^{CDa}			
6	NaCl+Sp7	0.84±0.37^{Bb}	0.00±0.00^{Aa}	0.00±0.00^{Ac}			
7	NaCl+NH+U	4.98±0.14^{Fb}	4.92±0.27^{Ea}	4.92±0.27^{Db}			
8	NaCl+Sp7+U	3.26±0.44^{Cb}	0.62±0.49^{Ba}	0.62±0.49^{Ba}			
9	Sp7	4.84±0.65^{DEFa}	4.84±0.65^{DE}	4.84±0.65^{Ba}			
10	NH	4.74±0.66^{Da}	4.74±0.66^{Da}	4.74±0.66^{BCa}			

A-H and a-c express the statistic evaluation and compare the 10 treatments (plots) (1-10) within each experiment (columns) and the three experiments within the same treatment (lines), respectively. Values accompanied by the same letter are not significantly different according to Fisher's Least Significant Difference test ($P<0.05$).

Table 3. Evaluation of the wheat growth under the various treatments

N°	Treatments	SH (Stem height) (cm)			HgL (Leaf height) (cm)		
		160 mM NaCl	200 mM NaCl	Field (200) mM NaCl)	160 mM NaCl	200 mM NaCl	Field (200) mM NaCl
1	Control	60.20±0.50^{Ea}	60.20±0.50^{Ea}	66.10±0.40^{Gb}	16.70±0.71^{Ca}	25.10±0.32^{Fa}	27.5±0.20^{Fb}
2	U	65.10±2.43^{Ha}	65.10±2.43^{Ga}	67.80±0.61^{Hb}	25.10±0.32^{Ga}	25.10±0.20^{Ba}	28.3±0.30^{Gb}
3	NaCl	37.60±0.50^{Ab}	0.00±0.00A^{a}	0.00±0.00^{Aa}	15.70±0.24^{Ab}	0.00±0.00^{Aa}	0.00±0.00^{Aa}
4	NaCl+U	46.90±0.80^{Bc}	30.30±4.83^{Ba}	41.50±1.91^{Cb}	19.50±0.34^{Dc}	12.30±2.02^{Ca}	17.3±0.80^{Cb}
5	NaCl+NH	51.70±1.73^{Dc}	38.60±0.61^{Ca}	45.40±2.60^{Db}	21.50±0.72^{Fc}	16.04±0.30^{Da}	18.9±1.10^{Db}
6	NaCl+Sp7	38.01±2.31^{Ab}	0.00±0.00^{Aa}	0.00±0.00^{Aa}	15.90±1.30^{Bb}	0.00±0.00^{Aa}	0.00±0.00^{Aa}
7	NaCl+NH+U	61.70±0.53^{FGc}	45.90±0.40^{Da}	47.20±1.99^{Eb}	25.70±0.30^{Ic}	19.04±0.21^{Ea}	19.6±1.10^{Eb}
8	NaCl+Sp7+U	48.70±1.01^{Cc}	30.10±5.72^{Ba}	38.10±3.98^{Bb}	20.30±0.41^{Ec}	12.30±2.46^{Ca}	15.99±1.90^{Bb}
9	Sp7	62.10±0.53^{Ga}	62.10±0.53^{Fa}	69.80±0.51^{Ib}	25.90±0.41^{Ia}	25.90±0.26^{Ga}	29.10±0.23^{Hb}
10	NH	61.20±0.81^{Fa}	61.30±0.81^{Fa}	64.20±0.35^{Fb}	25.40±0.42^{Ha}	25.40±0.42^{Fa}	28.8±0.20^{Hb}

N°	Treatments	WdL (Leaf width) (mm)			DWS (Stem dry weight) (g)		
		160 mM NaCl	200 mM NaCl	Field (200) mM NaCl	160 mM NaCl	200 mM NaCl	Field (200) mM NaCl
1	Control	6.22±0.20^{Da}	7.50±0.65^{Ea}	11.00±0.00^{Eb}	1.98±0.03^{Fa}	1.98±0.03^{Fa}	2.21±0.03^{Eb}
2	U	7.50±0.65^{Fa}	7.50±0.65^{Ga}	11.00±0.00^{Eb}	2.17±0.10^{Ia}	2.17±0.10^{Ia}	2.28±0.02^{Fb}
3	NaCl	5.00±0.00^{Bb}	0.00±0.00^{Aa}	0.00±0.00^{Aa}	0.30±0.02^{Ab}	0.00±0.00^{Aa}	0.00±0.00^{Aa}
4	NaCl+U	5.98±0.14^{Cc}	3.64±0.70^{Ba}	4.93±0.25^{Bb}	0.44±0.01^{Cc}	0.20±0.04^{Ca}	0.29±0.02^{Bb}
5	NaCl+NH	6.42±0.54^{Ec}	5.00±0.00^{Ca}	5.90±0.40^{Cb}	1.27±0.05^{Ec}	0.31±0.01^{Da}	0.33±0.02^{Cb}
6	NaCl+Sp7	4.86±0.45^{Ab}	0.00±0.00^{Aa}	0.00±0.00^{Aa}	0.32±0.03^{Bb}	0.00±0.00^{Aa}	0.00±0.00^{Aa}
7	NaCl+NH+U	8.00±0.00^{Gc}	6.00±0.00^{Da}	6.067±0.25^{Db}	2.04±0.03^{Hc}	0.41±0.01^{Eb}	0.35±0.02^{Da}
8	NaCl+Sp7+U	6.00±0.00^{Cc}	3.50±0.73^{Ba}	4.967±0.50^{Bb}	0.47±0.01^{Dc}	0.18±0.05^{Ba}	0.28±0.04^{Bb}
9	Sp7	8.00±0.00^{Ga}	8.00±0.00^{Fa}	11.57±0.50^{Fb}	2.04±0.03^{Ha}	2.04±0.03^{Ha}	2.30±0.01^{Gb}
10	NH	7.96±0.20^{Ga}	7.96±0.20^{Fa}	11.13±0.35^{Eb}	2.02±0.02^{Ga}	2.02±0.02^{Ga}	2.28±0.01^{Fb}

N°	Treatments	DWR (Roots dry weight) (g)			TLS (Total spike length, with hairs) (cm)		
		160 mM NaCl	200 mM NaCl	Field (200) mM NaCl	160 mM NaCl	200 mM NaCl	Field (200) mM NaCl
1	Control	0.49±0.01^{Fa}	0.49±0.01^{Fa}	0.64±0.01^{Eb}	15.05±0.12^{Ea}	15.49±0.12^{Fa}	17.68±0.28^{Fb}
2	U	0.53±0.02^{Ha}	0.53±0.02^{Ga}	0.66±0.01^{Fb}	17.14±0.93^{Ha}	17.14±0.92^{Fa}	18.45±0.20^{Gb}
3	NaCl	0.12±0.01^{Ab}	0.00±0.00^{Aa}	0.00±0.00^{Aa}	10.14±0.43^{Ab}	0.00±0.00^{Aa}	0.00±0.00^{Aa}
4	NaCl+U	0.19±0.01^{Cc}	0.08±0.01^{Ca}	0.10±0.01^{Cb}	11.92±0.40^{Bb}	8.57±13.25^{Ca}	12.51±0.90^{Cb}
5	NaCl+NH	0.27±0.02^{Ec}	0.14±0.01^{Db}	0.12±0.01^{Da}	13.37±0.51^{Dc}	10.56±0.17^{Da}	12.80±0.63^{Db}
6	NaCl+Sp7	0.14±0.01^{Bb}	0.00±0.00^{Aa}	0.00±0.00^{Aa}	10.17±1.20^{Ab}	0.00±0.00^{Aa}	0.00±0.00^{Aa}
7	NaCl+NH+U	0.52±0.01^{Gc}	0.17±0.02^{Eb}	0.12±0.01^{Da}	15.83±0.31^{Fc}	12.92±0.25^{Ea}	13.41±0.43^{Eb}
8	NaCl+Sp7+U	0.21±0.10^{Db}	0.07±0.02^{Ba}	0.09±0.01^{Ba}	12.36±0.41^{Cb}	6.62±1.80^{Ba}	11.99±0.80^{Bb}
9	Sp7	0.62±0.02^{Ja}	0.62±0.02^{Ia}	0.69±0.01^{Hb}	15.66±0.33^{EFa}	15.66±1.80^{Fa}	18.86±0.30^{Hb}
10	NH	0.61±0.01^{Ia}	0.61±0.01^{Ha}	0.68±0.01^{Gb}	16.10±0.22^{Ga}	16.10±0.22^{Fa}	18.59±0.24^{Gb}

N°	Treatments	LH (Hairs length) (cm)			LS (Spike length without hairs) (cm)		
		160 mM NaCl	200 mM NaCl	Field (200) mM NaCl	160 mM NaCl	200 mM NaCl	Field (200) mM NaCl
1	Control	9.05±0.08^{DEa}	9.05±0.08^{Da}	10.75±0.30^{Eb}	6.44±0.13^{Ea}	6.44±0.13^{Da}	6.93±0.06^{Fb}
2	U	10.41±0.6^{Ha}	10.41±0.10^{Fa}	11.43±0.15^{Fb}	6.73±0.50^{Fa}	6.73±0.50^{Da}	7.01±0.06^{FGb}
3	NaCl	7.12±0.40^{Ab}	0.00±0.00^{Aa}	0.00±0.00^{Aa}	3.02±0.10^{Ab}	0.00±0.00^{Aa}	0.00±0.00^{Aa}
4	NaCl+U	7.98±0.40^{Bb}	4.90±1.02^{Ba}	8.98±0.75^{Cc}	3.94±0.40^{Ba}	3.71±13.11^{Ca}	3.53±0.20^{Ca}
5	NaCl+NH	9.00±0.40^{Db}	7.53±0.13^{Ca}	9.12±0.40^{Cb}	4.40±0.25^{Dc}	3.03±0.10B^{Ca}	3.70±0.32^{Db}
6	NaCl+Sp7	7.14±0.80^{Ab}	0.00±0.00^{Aa}	0.00±0.00^{Aa}	3.03±0.50^{Ab}	0.00±0.00^{Aa}	0.00±0.00^{Aa}
7	NaCl+NH+U	9.34±0.25^{Fa}	9.44±0.23^{Ea}	9.35±0.30^{Da}	6.50±0.10^{Ec}	3.50±0.12^{Ca}	4.06±0.20^{Eb}
8	NaCl+Sp7+U	8.30±0.36^{Cb}	4.90±1.40^{Ba}	8.68±0.64^{Bb}	4.05±0.10^{Cc}	1.80±0.45^{Ba}	3.31±0.24^{Bb}
9	Sp7	9.20±0.26^{Ea}	9.20±0.30^{Da}	11.69±0.20^{Gb}	6.50±0.11^{Ea}	6.50±0.11^{Da}	7.20±0.20^{Hb}
10	NH	9.60±0.20^{Ga}	9.60±0.20^{Ea}	11.54±0.20^{FGb}	6.52±0.12^{Ea}	6.52±0.12^{Da}	7.05±0.11^{Gb}

N°	Treatments	DWS (Spike dry weight) (g)			SdNS (Seeds number per spike)		
		160 mM NaCl	200 mM NaCl	Field (200) mM NaCl	160 mM NaCl	200 mM NaCl	Field (200) mM NaCl
1	Control	2.70±0.02^{Ea}	2.70±0.02^{Fa}	2.91±0.05^{Fb}	35.80±0.70^{Fa}	35.80±10^{Fa}	38.00±0.00^{Fb}
2	U	2.90±0.10^{Ga}	2.90±0.10^{Ha}	2.97±0.02^{Fb}	37.96±0.30^{Ga}	37.96±0.30^{Ga}	38.00±0.00^{Fa}
3	NaCl	1.60±0.03^{Bb}	0.00±0.00^{Aa}	0.00±0.00^{Aa}	19.92±0.40^{Bb}	0.00±0.00^{Aa}	0.00±0.000^{Aa}
4	NaCl+U	2.10±0.10^{Cc}	1.10±0.30^{Ba}	1.90±0.12^{Cb}	25.50±1.20^{Cc}	12.92±2.50^{Ba}	22.70±2.10^{Cb}
5	NaCl+NH	2.40±0.10^{Dc}	1.60±0.02^{Da}	2.15±0.20^{Db}	29.60±0.93^{Ec}	18.00±0.00^{Da}	25.60±1.92^{Db}
6	NaCl+Sp7	1.45±0.20^{Ab}	0.00±0.00^{Aa}	0.00±0.00^{Aa}	17.60±1.80^{Ab}	0.00±0.00^{Aa}	0.00±0.00^{Aa}
7	NaCl+NH+U	2.80±0.05^{Fc}	2.12±0.01^{Ea}	2.22±0.10^{Eb}	36.00±0.00^{Fc}	26.00±0.00^{Ea}	26.80±1.54^{Eb}
8	NaCl+Sp7+U	2.11±0.05^{Cc}	1.15±0.25^{Ca}	1.70±0.30^{Bb}	25.96±0.30^{Dc}	14.04±2.93^{Ca}	19.93±3.40^{Bb}
9	Sp7	2.92=0.10^{Ha}	2.92±0.10^{Ha}	3.20±0.02^{Gb}	36.00±0.00^{Fa}	36.00±0.00^{Fa}	38.00±0.00^{Fb}
10	NH	2.80±0.11^{Fa}	2.80±0.11^{Ga}	3.13±0.20^{Gb}	36.00±0.00^{Fa}	36.00±0.00^{Fa}	38.00±0.00^{Fb}

N°	Treatments	TDWSdS (Total weight of seeds per spike) (g)			1000-seeds weight (g)		
		160 mM NaCl	200 mM NaCl	Field (200) mM NaCl	160 mM NaCl	200 mM NaCl	Field (200) mM NaCl
1	Control	2.30±0.02^{Fa}	2.30±0.02^{Fa}	2.45±0.02^{Fb}	64.22±1.23^{Ba}	64.22±1.23^{Da}	64.51±0.70B^{Ca}
2	U	2.40±0.05^{Ha}	2.40±0.05^{Ha}	2.50±0.01^{Fb}	63.12±1.22^{Aa}	63.12±1.22^{Ca}	65.20±0.30^{EFb}
3	NaCl	1.30±0.03^{Bb}	0.00±0.00^{Aa}	0.00±0.00^{Aa}	64.70±0.73^{Bb}	00.00±00.00^{Aa}	00.00±00.00^{Aa}
4	NaCl+U	1.70±0.09^{Cc}	0.80±0.16^{Ba}	1.50±0.10^{Cb}	65.50±1.10^{CDc}	60.30±2.30^{Ba}	64.60±2.20^{BCDb}
5	NaCl+NH	1.95±0.08^{Ec}	1.20±0.03^{Da}	1.70±0.22^{Db}	66.10±1.54^{Da}	64.90±1.55^{EFa}	64.81±7.02^{CDEa}
6	NaCl+Sp7	1.21±0.15^{Ab}	0.00±0.00^{Aa}	0.00±0.00^{Aa}	68.80±2.80^{Fb}	00.00±00.00^{Aa}	00.00±00.00^{Aa}
7	NaCl+NH+U	2.33±0.02^{Gc}	1.70±0.02^{Ea}	1.80±0.09^{Eb}	64.74±0.50^{Ba}	64.44±0.80^{DEa}	65.84±1.22^{Gb}
8	NaCl+Sp7+U	1.70±0.05^{Dc}	0.90±0.20^{Ca}	1.30±0.23^{Bb}	64.74±1.97^{CDc}	62.90±1.94^{Ca}	64.30±1.11^{Bb}
9	Sp7	2.43±0.04^{Ia}	2.43±0.04^{Ia}	2.50±0.01^{Fb}	67.53±0.99^{Eb}	67.53±0.99^{Gb}	65.45±0.34^{FGa}
10	NH	2.35±0.07^{Ga}	2.35±0.10^{Ga}	2.50±0.02^{Fb}	65.40±1.99^{Ca}	65.40±1.99^{Fa}	65.10±0.50^{DEFa}

N°	Treatments	CHA (OD of chlorophyll a)			CHB (OD of chlorophyll b)		
		160 mM NaCl	200 mM NaCl	Field (200) mM NaCl	160 mM NaCl	200 mM NaCl	Field (200) mM NaCl
1	Control	2.50±0.02^{Ea}	2.50±0.02^{Fa}	2.80±0.03^{Eb}	2.30±0.02^{Ea}	2.30±0.02^{Fa}	2.60±0.03^{Eb}
2	U	2.71±0.05^{Ha}	2.71±0.05^{Ha}	2.83±0.03^{EFb}	2.51±0.06^{Ha}	2.51±0.06^{Ha}	2.62±0.03^{EFb}
3	NaCl	1.60±0.02^{Ab}	0.00±0.00^{Aa}	0.00±0.00^{Aa}	1.35±0.03^{Ab}	0.00±0.00^{Aa}	0.00±0.00^{Aa}
4	NaCl+U	1.95±0.03^{Bc}	1.30±0.08^{Ca}	1.70±0.12^{Bb}	1.74±0.04^{Bc}	1.10±0.08^{Ca}	1.50±0.13^{Bb}
5	NaCl+NH	2.13±0.20^{Dc}	1.60±0.05^{Da}	1.90±0.21^{Cb}	1.92±0.20^{Dc}	1.20±0.04^{Da}	1.70±0.22^{Cb}
6	NaCl+Sp7	1.61±0.06^{Ab}	0.00±0.00^{Aa}	0.00±0.00^{Aa}	1.41±0.10^{Ab}	0.00±0.00^{Aa}	0.00±0.00^{Aa}
7	NaCl+NH+U	2.60±0.01^{FGc}	1.91±0.01^{Ea}	2.15±0.21^{Db}	2.40±0.02^{FGc}	1.71±0.02^{Ea}	1.94±0.21^{Db}
8	NaCl+Sp7+U	2.02±0.03^{Cc}	1.20±0.21^{Ba}	1.70±0.30^{Bb}	1.82±0.02^{Cc}	0.98±0.21^{Ba}	1.50±0.30^{Bb}
9	Sp7	2.60±0.01^{Ga}	2.60±0.01^{Ga}	2.90±0.04^{Fb}	2.40±0.02^{Ga}	2.40±0.02^{Ga}	2.70±0.04^{Fb}
10	NH	2.54±0.03^{EFa}	2.54±0.03^{FGa}	2.90±0.02^{Fb}	2.34±0.03^{EFa}	2.34±0.03^{FGa}	2.70±0.04^{EFb}

N°	Treatments	CHT (Total rate in chlorophyll) (mg/g of fresh matter)			PRO (Proline content) (µg/g of fresh matter)		
		160 mM NaCl	200 mM NaCl	Field (200) mM NaCl	160 mM NaCl	200 mM NaCl	Field (200) mM NaCl
1	Control	6.65±0.10^{Ea}	6.65±0.07^{Fa}	7.42±0.10^{Eb}	2.10±0.11^{Ab}	2.10±0.11^{Ab}	1.90±0.04^{Ba}
2	U	7.24±0.20^{Ha}	7.24±0.20^{Ha}	7.55±0.10^{EFb}	1.99±0.05^{Ab}	1.99±0.05^{Ab}	1.80±0.06^{Ba}
3	NaCl	3.99±0.10^{Ab}	0.00±0.00^{Aa}	0.00±0.00^{Aa}	78.92±1.30^{Db}	0.00±0.00^{Aa}	0.00±0.00^{Aa}
4	NaCl+U	5.10±0.10^{Bc}	3.25±0.22^{Ca}	4.40±0.40^{Bb}	57.20±1.93^{Ca}	92.99±4.30^{Dc}	67.70±0.6^{Eb}
5	NaCl+NH	5.60±0.50^{D}c	4.11±0.13^{Da}	4.90±0.60^{Cb}	49.20±14.92^{Ba}	75.71±2.10^{Cc}	57.50±0.85^{Db}
6	NaCl+Sp7	4.13±0.20^{Ab}	0.00±0.00^{Aa}	0.00±0.00^{Aa}	75.33±2.90^{Db}	0.00±0.00^{Aa}	0.00±0.00^{Aa}
7	NaCl+NH+U	6.90±0.04^{FGc}	4.99±0.05^{Ea}	5.64±0.60^{Db}	2.02±0.01^{Aa}	58.75±0.75^{Bc}	51.50±0.60^{Cb}
8	NaCl+Sp7+U	5.30±0.10^{Cc}	2.84±0.70^{Ba}	4.23±0.90^{Bb}	52.91±1.72^{BCa}	98.84±11.65^{Ec}	67.93±0.42^{Eb}
9	Sp7	6.93±0.05^{Ga}	6.93±0.05^{Ga}	7.80±0.11^{Fb}	2.01±0.01^{Ab}	2.01±11.65^{Ab}	1.74±0.02^{Ba}
10	NH	6.80±0.10^{EFa}	6.80±0.10^{FGa}	7.75±0.05^{EFb}	2.03±0.05^{Ab}	2.03±0.05^{Ab}	1.80±0.02^{Ba}

N°	Treatments	TS (Total sugars rate) (µg/g of fresh matter)		
		160 mM NaCl	200 mM NaCl	Field (200) mM NaCl
1	Control	28.80±0.98^{Ab}	28.80±0.98^{Cb}	19.70±0.33^{Ba}
2	U	25.98±0.80^{Ab}	25.98±0.76^{BCb}	18.92±0.20^{Ba}
3	NaCl	252.90±3.93^{Db}	0.00±0.00^{Aa}	0.00±0.00^{Aa}
4	NaCl+U	183.20±6.23^{Ca}	306.83±10.74^{FC}	214.40±2.63^{Eb}
5	NaCl+NH	157.45±47.99^{Bb}	242.02±6.22^{Eb}	183.85±2.71^{Da}
6	NaCl+Sp7	240.90±9.62^{Db}	0.00±0.00^{Aa}	0.00±0.00^{Aa}
7	NaCl+NH+U	38.50±57.11^{Aa}	187.80±2.75^{Dc}	163.32±1.23^{Cb}
8	NaCl+Sp7+U	164.80±7.40^{BCa}	317.31±19.45^{Gc}	217.91±2.04^{Fb}
9	Sp7	19.85±0.33^{Ab}	19.85±0.33^{Bb}	19.30±0.55^{Ba}
10	NH	20.05±0.15^{Ab}	20.05±0.15^{Bb}	19.24±0.41^{Ba}

The application of algal extract increased chlorophyll content of wheat plants in the absence of saline stress (Table 3). *U. lactuca* extract probably provides plants with nutritive elements enhancing photosynthesis. The same results were obtained with *A. brasilense* NH and Sp7 in the absence of salt. The effect of these two bacteria on the chlorophyll pigmentation could be due to an improved mineral and nitrogen as well as water uptake. The partial recovery of the chlorophyll content from markedly reduced chlorophyll levels at NaCl-stress conditions in the presence of *A. brasilense* NH or algal extracts and the complete recovery up to the control level upon the combined application of *A. brasilense* NH and algal extracts was probably due to the concerted actions of several different beneficial effects. *A. brasilense* NH restored the sufficient water and nitrogen content and the algal extracts provided the compatible solutes which reduced the inhibitory effect of increased salinity as well as several nutrients. The adaptation of plants to osmotic stress is associated with metabolic adjustments leading e.g. to the accumulation of organic compatibles solutes such as sugars, polyols, betaines and proline. Proline accumulation in plant tissues is an important physiological response to counterbalance saline stress. The greatly increased proline levels found in wheat plants during severe salt stress (Table 3) are reflecting this response. The inoculation of salt-stressed wheat with *A. brasilense* NH or the addition of algal extracts effectively reduced the proline content (Table 2). Since proline is a potent osmolyte accumulated under salt stress, this indicated reduced salt stress. Aqueous extract of *U. lactuca* provides a proline source, various betaines and DMSP [24] and other compatible solutes. Thus, under saline stress conditions, the wheat plants could directly use the compatible solutes available in the root environment by the addition of algal extracts for osmotic adjustment. Salt induced proline accumulation in the plants is reversed providing again various amino acids necessary for growth and development [62]. The contents of proline and chlorophyll measured in our study agree with those of other authors [63] which suggested a possible connection between chlorophyll pigment biosynthesis pathways and the proline content. The accumulation of sugars is another important adaptive response of plants and other organisms to the lack of water and saline stress. Sugars play an important role during germination of seeds under osmotic stress [64]. The accumulation of sugars in salinity stressed plants prevents structural and functional changes of membranes and destruction of soluble proteins. Soluble sugars (trehalose, mannitol, glucose and fructose) are known for their osmoprotective effect in many species of wheat (trehalose, mannitol, glucose and fructose) [65, 66]. Our data of high total sugar contents at most pronounced salt stressed conditions (treatments 3 and 6, Table 3) coincides with this general response. Accordingly, the sugar content is reduced in treatments 4, 5, 7 and 8, when either algal extracts or *A. brasilense* NH or both were applied and osmotic stress is apparently reduced. The sugar content was found to respond to the experimental treatments exactly like the proline content (Fig. 2). Saline stressed *T. aestivum* L. plants responded also with increases in various sugar compounds [67].

The improvement and the restoration of wheat growth under saline conditions by *A. brasilense* NH could be in part due to the IAA-production of this bacterium performing optimally at 200 mM NaCl [23]. IAA is the most important phytohormone produced by *A. brasilense*, which was shown to be responsible for plant growth promotion from the germination to the harvest stages and for morphology changes of roots, improved shoot growth and improved yields when present at optimal concentrations [68, 69, 70]. The levels of phytohormones are crucial for the protection of the plant against various stresses. It was reported that salinity stress resulted in a progressive decline of IAA-levels in wheat plants [71]. The ability of *A. brasilense* NH to produce IAA at salinity levels could complement this decrease in the IAA-levels of the roots and thus restore plant performance at salinity stress. In addition, *Azospirillum brasilense* was demonstrated to produce exopolysaccharides (EPS) under saline stress conditions [72]. These EPS are released into the rhizosphere and form the EPS-cation-soil complex which is able to neutralize Na+ ions in soil and prevent their uptake by the plant. This complex is also able to restore the water content in soil [73, 74, 75]. Another important mechanism of plant growth stimulation by rhizobacteria is operating through the bacterial degradation of the precursor ACC of the phytohormone ethylene by ACC-deaminase (Mayak et al., 2004)[7]. Decreasing elevated ethylene levels in roots during salinity stress leads to improved plant development. ACC-deaminase is present in many PGPR [76, 77], but our data do not support the presence of ACC-deaminase activity in *A. brasilense* NH. Thus, ACC-deaminase seems not to be involved in the acquirement of halotolerance in this case.

It is most remarkable, that FISH-analysis using species-specific fluorescently labeled oligonucleotide probes combined with confocal laser scanning microscopy [38] could demonstrate an abundant colonization of *A. brasilense* NH in the intercellular spaces of the wheat root cortex (Figure 3). Endophytic bacteria were shown in several cases to exert most interesting effects on plants, leading to efficient biological control, nitrogen fixation and plant growth promotion [78, 79]. Within the species *Azospirillum brasilense*, a similar endophytic colonization is known for the strain Sp245, which also colonize the root cortex of wheat plants [80, 81]. Endophytic colonization of plants by microbes is only possible when the plant is not recognizing the bacterial cells as enemy and mobilizes defense reactions; on the other hand, the bacterium has abilities to enter and thrive within a plant successfully, exerting its beneficial effects with less competition with the abundant rhizosphere microflora. The observed apparently symbiotic interaction of the halotolerant and slightly halophilic strain NH with wheat roots may be regarded as a result of coevolution under saline conditions [82] which prevails since thousands of years in soils close to the Mediterranean coast of Algeria.

CONCLUSION

The successful restoration of growth of durum wheat under salinity conditions after inoculation with *A. brasilense* NH with and without concomitant application of aqueous extracts of *U. lactuca* – a natural resource from the sea coast - provides the basis for a new approach of a successful formulation of a bacterial seed inoculum for the improvement of growth of durum wheat in saline soil.

ACKNOWLEDGMENT

We kindly appreciate the excellent technical assistance of Mrs. Angelika Schulz (Helmholtz Zentrum München, Germany) in the laboratory experiments with FISH-analysis. Furthermore, many thanks to Benyahia Fateh (INRA Company) for the great help in the performance of the pot and field experiments in Bejaïa, and to Dr. M. Sahnoune for the plant analysis.

REFERENCES

[1] Carpita, N. C. (1985). Tensile strength of cell walls of living cells. *Plant Physiol*, 79, 485-488.

[2] Cordovilla, M. P., Ligero, F. & Lluch, C. (1994). The effect of salinity on N-fixation on growth, symbiotic performance and nitrogen assimilation in Faba bean (*Vicia faba* L.) under salt stress. *Plant Soil*, 172, 289-297.

[3] Zahran, H. H. (1999). *Rhizobium*-legume symbiosis and nitrogen fixation under severe conditions and in an arid climate. *Microbiol Mol Biol Rev*, 63, 968-989.

[4] Bartels, D. & Sunkar, R. (2005). Drought and Salt Tolerance in Plants. *Critical Reviews in Plant Sciences*, 24, 23-58.

[5] Creus, C. M., Sueldo, R. J, Barassi, C. A. (1998). Water relations in *Azospirillum*-inoculated wheat seedlings under osmotic stress. *Can J Bot,* 76, 238-244.

[6] Egamberdiyeva, D. (2005). Characterization of Pseudomonas species isolated from the rhizosphere of plant s grown in serozem soil, semi arid region of Uzbekistan. *Scientific World Journal*, 5, 501-509.

[7] Mayak, S, Tirosh, T. Glick, B. R. (2004). Plant growth-promoting bacteria that confer resistance in tomato to salt stress. *Plant Physiol Biochem*, 42, 565-572.

[8] Ghoul, M. L'halotolérance de *E. coli*. *Effet des osmoprotecteurs naturels.* Doctorate thesis Université of Rennes I. France. (1990).

[9] Epstein, W. (1986). Osmoregulation by potassium transport in *Escherichia coli. FEMS Microbiol Lett*, 39, 73-78.

[10] Bakker, E. P., Booth, I. R., Dinnbier, U., Epstein, W. & Gajewska, A. (1987). Evidence for multiple K^+ export systems in *Escherichia coli. J Bacteriol,* 169, 3743-3749.

[11] Kogut, M. & Russell, N. J. (1987). Life at the limits: Considerations on how bacteria can grow at extremes of temperature and pressure or with high concentrations of ions and solutes. *Sci Prog*, 71, 381-400.

[12] Jofré, E, Fischer, S., Rivarola, V., Balegno, H. & Mori, G. (1998). Saline stress affects the attachment of *Azospirillum brasilense* Cd to maize and wheat roots. *Can J Microbiol*, 44, 416-422.

[13] Phillips, D. A., Joseph, C. M. & Maxwell, C. A. (1992). Trigonelline and stachydrine released from alfalfa seeds activate NodD2 protein in *Rhizobium meliloti. Plant Physiol*, 99, 1526-1531.

[14] Xiong, L. & Zhu, J. K. (2002). Molecular and genetic aspects of plant responses to osmotic stress. *Plant Cell Environ*, 25, 131-139.

[15] Dobbelaere, S., Croonenborghs, A., Amber, T., Ptacek, D., Vanderleyden, J., Dutto, P., Labandera-Gonzalez, C., Caballero-Mellado, J., Aguirre, J. F., Kapulnik, Y., Shimon, B., Burdman, S., Kadouri, D., Sarig, S. & Okon, Y. (2001). Responses of agronomically important crops to inoculation with *Azospirillum. Aust J Plant Physiol*, 28, 1-9.

[16] Hartmann, A. & Baldani, J. I. (2006). The genus *Azospirillum*. In: M. Dworkin, S. Falkow, E. Rosenberg, K. H. Schleifer. & E. Stackebrandt (Eds.), *The Prokaryotes vol. 5 : Proteobacteria, Alpha and Beta Subclasses* (Springer, Chapter 3.1.5, pp. 115-140). New York, USA.

[17] Okon, Y. & Kapulnik, Y. (1986). Development and function of *Azospirillum*-inoculated roots. *Plant Soil,* 90, 3-16.

[18] Okon, Y. & Labandera-Gonzalez, C. (1994). Agronomic applications of *Azospirillum*: An evaluation of 20 years world-wide field inoculation. *Soil Biol Biochem,* 26, 1591-1601.

[19] Okon, Y. & Vanderleyden, J. (1997). Root-associated *Azospirillum* species can stimulate plants. *ASM News*, 63, 366-370.

[20] Bashan, Y. & Holguin, G. (1997). *Azospirillum*-plant relationships, environmental and physiological advances. *Can J Microbiol*, 43, 103-121.

[21] Dobbelaere, S., Vanderleyden, J. & Okon, Y. (2003). Plant growth-promoting effects of diazotrophs in the rhizosphere. *Critical Reviews in Plant Sciences,* 22, 107 - 149.

[22] Perrig, D., Boiero, M. L., Masciarelli, O. A., Penna, C., Ruiz, O. A., Cassan, F. D. & Luna, M. V. (2007). Plant-growth-promoting compounds produced by two agronomically important strains of Azospirillum brasilense, and implications for inoculant formulation. *Appl Microbiol Biotechno*l, 75, 1143-1150.

[23] Nabti, E., Sahnoune, M., Adjrad, S., Van Dommelen, A., Ghoul, M., Schmid, M. & Hartmann, A. (2007). A halophilic and osmotolerant *Azospirillum brasilense* strain from algerian soil restores wheat growth under saline conditions. *Eng Life Sci*, 7, 354-360.

[24] Ghoul, M., Minet, J., Bernard, T., Dupray, E. & Cormier, M. (1995). Marine macro-algae as a source for osmoprotection for *Escherichia coli*. *Microb Ecol*, 30, 171-181.

[25] Galinski, E. A. (1995) Osmoadaptation in bacteria. *Adv Microb Physiol,* 37:274-325.

[26] Pichereau, V., Pocard, J. A., Hamelin, J., Blanco, C. & Bernard, T. (1998). Differential effects of dimethylsulfoniopropionate, dimethylsulfonioacetate, and other S-methylated compounds on the growth of *Sinorhizobium meliloti* at low and high osmolarities. *Appl Environ Microbiol*, 64, 1420-1429.

[27] Malmstrom, R. R., Kiene, R. P., Cottrell, M. T. & Kirchman, D. L. (2004). Contribution of SAR11 bacteria to dissolved dimethylsulfoniopropionate and amino acid uptake in the north atlantic ocean. *Appl Environ Microbiol*, 70, 4129-4135.

[28] Reed, R. H. (1983). Measurement and osmotic significance of ß-dimethylsulphonio-propionate in marine macroalgae. *Mar Biol Lett,* 34,173-181.

[29] Rhodes, D. & Hanson, A. D. (1993). Quaternary ammonium and tertiary sulfonium compounds in higher plants. *Annu Rev Plant Physiol Plant Mol Biol,* 44, 357-384.

[30] Saqib, M. Akhtar, J., Qureshi, R. H. (2004). Pot study on wheat growth in saline and waterlogged compacted soil: I grain yield and yield components. *Soil Till Res*, 77, 169-177.

[31] Sepaskhah, A. R., Bazrafshan-Jahromi, A. R. & Shirmohammadi-Aliakbarkhani, Z. (2006). Development and evaluation of a model for yield production of wheat, maize and sugarbeet under water and salt stresses. *Biosyst Eng*, 93:139-152.

[32] Cayla, M.. *Découvrez les algues: propriétés, applications, recettes*. Paris, France. Chairon; 1995; 23-25.

[33] STAT-ITCF (1992). *Logiciels et manuels d'utilisation*. ITCF, Paris, France. STAT-ITCF. STATSOFT, 2004. Statistica, Version 7.0. StatSoft, Inc., Tulsa, Oklahoma.

[34] Bengston, C., Klockare, B., Klockare, R., Larsson, S. & Sudquist, C. (1978). The after effect of water stress on chlorophyll formation during greening and the level of abscisic acid and proline in dark growth wheat seedlings. *Plant Physiol*, 43, 205-212.

[35] Aßmus, B., Hutzler, P., Kirchhof, G., Amann, R., Lawrence, J. R. & Hartmann, A. (1995). In situ localization of *Azospirillum brasilense* in the rhizosphere of wheat with fluorescently labeled, rRNA-targeted oligonucleotide probes and scanning confocal laser microscopy. *Appl Environ Microbiol,* 61, 1013-1019.

[36] Amann, R. I., Binder, B. J., Olson, R. J., Chisholm, S. W., Devereux, R. & Stahl, D. A. (1990). Combination of 16S rRNA-targeted oligonucleotide probes with flow cytometry for analyzing mixed microbial populations. *Appl Environ Microbiol*, 56, 1919-1925.

[37] Daims, H., Bruhl, A., Amann, R., Schleifer, K. H. & Wagner, M. (1999). The domain-specific probe EUB338 is insufficient for the detection of all bacteria: development and evaluation of a more comprehensive probe set. *Syst Appl Microbiol*, 22, 434-444.

[38] Stoffels, M., Castellanos, T. & Hartmann, A. (2001). Design and application of new 16S rRNA-targeted oligonucleotide probes for the *Azospirillum-Skermanella-Rhodocista*-cluster. *Syst Appl Microbiol*, 24, 83-97.

[39] Brown, C. M., Dilworth, M. J. (1975). Ammonia assimilation by *Rhizobium* cultures and bacteriods. *J Gen Microbiol*, 122, 61-67.

[40] Alvey, S., Yang, C. H., Buerkert, D., Crowley, D. E. (2003). Cereal/legume rotation effects on rhizosphere bacterial community structure in West African soils. *Biol Fert Soils*, 37, 73-82.

[41] Nedzarek, A. & Rakusa-Suszczewski, S. (2004). Decomposition of macroalgae and the release of nutrient in Admiralty Bay, King George Island, Antarctica. *Polar Biosci*, 17, 26-35.

[42] Hartmann, A., Singh, M. & Klingmüller, W. (1983). Isolation and characterization of *Azospirillum* mutants excreting high amounts of indole acetic acid. *Can J Microbiol*, 916-923.

[43] Kapulnik, Y., Okon, Y. & Henis, Y. (1985) Changes in root morphology of wheat caused by *Azospirillum* inoculation. *Can J Microbiol*, 31, 881–887.

[44] Kapulnik, Y., Okon, Y. & Henis, Y. (1987). Yield response of spring wheat (*Triticum aestivum*) to inoculation with *Azospirillum brasilense* under field conditions. *Biol Fert Soils,* 4, 27-35.

[45] Vande Broek, A., Lambrecht, M. & Vanderleyden, J. (1998). Bacterial chemotactic motility is important for the initiation of wheat root colonization by *Azospirillum brasilense*. *Microbiology*, 144, 2599-2606.

[46] Ferreira, M. C. B., Fernandes, M. S. & Döbereiner, J. (1987). Role of *Azospirillum brasilense* nitrate reductase in nitrate assimilation by wheat plants. *Biol Fert Soils*, 4, 47-53.

[47] Rodrigues, O., Didonet, A. D., Gouveia, J. A. & Soares, C. (2000). Nitrogen translocation in wheat inoculated with *Azospirillum* and fertilized with nitrogen. *Pesqui Agropecu Bras,* 35, 1473-1481.

[48] Tripathi, A. K., Nagarajan, T., Verma, S. C. & Le Rudulier, D. (2002). Inhibition of biosynthesis and activity of nitrogenase in *Azospirillum brasilense* Sp7 under salinity stress. *Curr Microbiol*, 44, 363–367.

[49] Rengasamy, P. (2002). Transient salinity and subsoil constraints to dryland farming in Australian sodic soils: an overview. *Aust J. Exp Agric,* 42, 351–361.

[50] Rengasamy, P., Chittleborough, D. & Helyar, K. (2003). Root-zone constraints and plant-based solutions for dryland salinity. *Plant Soil*, 257, 249-260.

[51] Abdel-Ghaffar, B. A., El-Shourbagy, M. N. & Basha, E. M. (1998). Responses of NaCl stressed wheat to IAA. *Proceedings of the 6th Egypt Botany Conference*, Cairo. Univ Giza Egypt, 6, 79-88.

[52] Gunes, A., Inal, A. & Alpaslan, M (1996). Effect of salinity on stomatal resistproline, and mineral composition of pepper. *Journal of Plant Nutrition*, 19, 389 - 396.

[53] Gallé, Á., Csiszár, J., Tari, I. & Erdei, L. (2002). Changes in water and chlorophyll fluorescence parameters under osmotic stress in wheat cultivars. *Acta Biol Szeged*, 46:85-86.

[54] Termaat, A., Passioura, J. B. & Munns, R. (1985). Shoot turgor does not limit shoot growth of NaCl-affected wheat and barley. *Plant Physiol*, 77, 869-872.

[55] Rashid, A. (1986). *Mechanism of salt tolerance in wheat* (*Triticum aestivum* L.). Ph.D. Thesis. Univ Agric Faisalabad Pakistan.

[56] Iqbal, R. & Mahmood, A. (1992). Structural studies on root nodules of *Leucaena leucocephola* with particular reference to the infection process. *Pak J Bot*, 24,142-152.

[57] Evans, L. T. (1993). *Crop Evolution, adaptation and yield* Cambridge University Press, Cambridge.

[58] Sayre, K. D., Rajaram, S. & Fischer, R. A. (1997). Yield potential progress in short bread wheats in northwest Mexico. *Crop Sci*, 37, 36-42.

[59] Sharma, R. C., Gupta, N. K., Gupta, S. & Hasegawa, H. (2005). Effect of NaCl salinity on photosynthetic rate, transpiration rate, and oxidative stress tolerance in contrasting wheat genotype. *Photosynthesis*, 43, 609-613.

[60] Francois, L. E., Maas, E. V., Donovan, T. J. & Youngs, V. L. (1986). Effect of salinity on grain yield and quality, vegetative growth, and germination of semi-dwarf and durum wheat. *Agron J*, 78, 1053-1058.

[61] Qurcshi, R. H., Ahmed, R., Ilyas, M. & Aslam, Z. (1980) Screening of wheat (Triticum aestivum L.) for salt tolerance. *Pak J Agric Sci*, 27, 9-26.

[62] Brenda, C., Fermin, T. S., Hahm, J. A., Radinsky, R. J., Kratochvil, J. E., Hall, Y. & Martin, L. (2005). Effect of proline and glutamine on the functional properties of wheat dough in winter wheat varieties. *Journal of Food Science,* 70, 273-278.

[63] Reddy, P. S. & Veeranjaneyulu, K. (1991). Proline metabolism in senescing leaves of hosgram (*Macrotyloma uniflorum* lam.). *Journal of Physiology*, 137, 381-383.

[64] Gill, P. K., Sharma, A. D., Singh, P., Bhullar, S. S. (2002). Osmotic stress-induced changes in germination, growth and soluble sugar content of *Sorghum bicolour* (L.) Moench seeds. *Bulg J Plant Physiol*, 28.

[65] Ashraf, M. Y., Khan, M. A. & Maqvi, S. S. M. (1991). Effect of salinity on seedling growth and solutes accumulation in two wheat genotypes. *Rachis*, 10, 30-31.

[66] El-Haddad, E. H. M. & O'Leary, J. W. (1994). Effect of salinity and K/Na ratio of irrigation water on growth and solute content of *Atriplex amnicola* and *Sorghum bicolor*. *Irr Sci*, 14, 127-133.

[67] Kerepesi, I. & Galiba, G. (2000). Osmotic and salt stress-induced alteration in soluble carbohydrate content in wheat seedlings. *Crop Sci,* 40, 482-487.

[68] Baldani, V. L. D. & Döbereiner, J. (1980). Host-plant specificity in the infection of cereals with *Azospirilum* spp. *Soil Biol Biochem*, 12, 433–439.

[69] Barbieri, P. & Galli, E. (1993). Effect on wheat root development of inoculation with an *Azospirillum brasilense* mutant with altered indole-3-acetic acid production. *Res Microbiol*, 144, 69-75.

[70] Bashan, Y. & Levanony, H. (1990). Current status of *Azospirillum* inoculation technology. *Azospirillum* as a challenge for agriculture. *Can J Microbiol*, 36, 591-607.

[71] Sakhabutdinova, A. R., Fatkhutdinova, D. R., Bezrukova, M. V. & Shakirova, F. M. (2003). Salicylic acid prevents the damaging action of stress factors on wheat plants. *Bulg J Plant Physiol*, 33, 314-319.

[72] Konnova, S. A., Makarov, O. E., Skvortsov, I. M. & Ignatov, V. V. (1994). Isolation, fractionation and some properties of polysaccharides produced in a bound form by *Azospirillum brasilense* and their possible involvement in *Azospirillum*-wheat root interactions. *FEMS Microbiol Lett,* 118, 93-99.

[73] Bashan, Y., Moreno, M. & Troyo, E. (2000). Growth promotion of the seawater-irrigated oilseed halophyte *Salicornia bigelovii* inoculated with mangrove rhizosphere bacteria and halotolerant *Azospirillum* spp. *Biol Fert Soils,* 32, 265-272.

[74] Marshall, K. C. (1975). Clay mineralogy in relation to survival of soil bacteria. *Annu Rev Phytopathol,* 13, 357-373.

[75] Morel, J. L., Habib, L., Plantureux, S. & Guckert, A. (1991). Influence of maize root mucilage on soil aggregate stability. *Plant Soil*, 136, 111-119.

[76] Blaha, D., Prigent-Combaret, C., Mirza, M. S. & Moënne-Loccoz, Y. (2006). Phylogeny of the 1-aminocyclopropane-1-carboxylic acid deaminase-encoding gene *acd*S in phytobeneficial and pathogenic Proteobacteria and relation with strain biogeography. *FEMS Microbiol Ecol,* 56, 455-470.

[77] Glick, B. R., Penrose, D. M. & Li J. (1998). A model for the lowering of plant ethylene concentrations by plant growth promoting bacteria. *Journal of Theoretical Biology*, 190, 63-68.

[78] Rosenblueth, M. & Martinez-Romero, E. (2006). Bacterial endophytes and their interactions with hosts. *Mol Plant Microbe Interact*, 19, 827-837.

[79] Schulz, B., Boyle, C. & Sieber, N. (2006*). Microbial root endophytes* (Springer) Verlag, Berlin, Heidelberg, New York.

[80] Rothballer, M., Schmid, M. & Hartmann, A. (2003). In situ localization and PGPR-effect of *Azospirillum brasilense* strains colonizing roots of different wheat varieties. *Symbiosis*, 34, 261-279.

[81] Schloter, M. & Hartmann, A. (1998). Endophytic and surface colonization of wheat roots (*Triticum aestivum*) by different *Azospirillum brasilense* strains studies with strain-specific monoclonal antibodies. *Symbiosis*, 25, 159-179.

[82] Zilber-Rosenberg, I. & Rosenberg, E. (2008). Role of microorganisms in the evolution of animals and plants: The hologenome theory of evolution. *FEMS Microbiol Rev*, 32, 723-735.

In: Algae
Editors: Dagmar Krueger and Helga Meyer

ISBN: 978-1-62081-280-8

Chapter 3

PRODUCTION AND PROPERTIES OF ALGAE OIL BIOFUEL AND ENVIRONMENTAL BENEFITS

Mohamed Y. E. Selim* and Yousef Haik
Mech. Eng. Dept., Faculty of Engineering,
UAE University, Al-Ain, United Arab Emirates

ABSTRACT

This chapter aims at testing the feasibility of developing an innovative microalgae base technology that can concurrently produce biodiesel and protect the environment. The use of fossil fuel for transportation systems is costing the world billions of dollars each year and contributes to air pollution and the global warming. Microalgae have been recognized as one of the most promising source of biofuel. However, there is no local testing or commercial production of microalgal oil in many countries so far.

In this work, it is hypothesized that an economically viable microalgae-based biodiesel production system could be established if innovative photobioreactors were used along with suitable microalgal species rich in oil content. It is proposed to combine the biodiesel production with wastewater treatment and CO_2 reduction using microalgae. Such development will commence commercial feasibility projects. This technology is particularly useful in countries, where huge amounts of CO_2 are released to environment and huge demand on useful applications of wastewater.

In all countries, billions of dollars are spent each year for energy production. Many of the power plants consume diesel in the process of generating electricity and desalination of water. Most public transportation system uses diesel engines. It is well established that carbon dioxide emission from various transportation systems accounts for one third of the total carbon dioxide released into the atmosphere, leading to global warming. The toxic exhausted gas from vehicles also causes hazardous effects on human health. Biodiesel, derived from vegetable oils or animal fats rather than petroleum, is an alternative fuel for diesel engines. It is renewable, non-toxic, and biodegradable. It can be used in existing diesel engines without modification, and can be blended in at any ratio with petroleum diesel. Currently, the major feedstock for biodiesel production is soybean oil. However, low oil production rates (e.g. 1 ~ 3 barrels of oil per acre of land per year)

* E-mail: mohamed.selim@uaeu.ac.ae.

coupled with high production costs (e.g. feedstock accounts for 70-80% of biodiesel production costs) have limited the expansion of soybean-based biodiesel to meet the growing demand by the society. Therefore, it is important to develop more economically viable sources of feedstock for biodiesel production. Microalgae are known to exhibit 10- to 20-fold higher growth rates than agricultural crop plants, and certain microalgal species can accumulate large amounts of lipids or oil (30-60% of dry weight).

In this work we propose to develop a complete prototyped model for the production and testing of biodiesel. The model will consist of cultivation reactor to grow suitable microalgae species, photobioreactors with wastewater and or CO_2 as feed to the algae, oil extraction from algae, chemical reactor to convert the extracted and finally test the produced biodiesel in a conventional diesel engine.

Keywords: Microalgae, biofuel, diesel, engine, performance

INTRODUCTION

Algae are the fastest-growing plants in the world. Industrial reactors for algal culture are open ponds, photo bioreactors and closed systems. Algae are very important as a biomass source. Different species of algae may be better suited for different types of fuel. Algae can be grown almost anywhere, even on sewage or salt water, and does not require fertile land or food crops, and processing requires less energy than the algae provides. Algae can be a replacement for oil based fuels, one that is more effective and has no disadvantages. Algae are among the fastest-growing plants in the world, and about 50% of their weight is oil. This lipid oil can be used to make biodiesel for cars, trucks, and airplanes [1].

Microalgae have much faster growth-rates than terrestrial crops. the per unit area yield of oil from algae is estimated to be from 20,000 to 80,000 l per acre, per year; this is 7–31 times greater than the next best crop, palm oil. The lipid and fatty acid contents of microalgae vary in accordance with culture conditions. Most current research on oil extraction is focused on microalgae to produce biodiesel from algal oil. Algal-oil processes into biodiesel as easily as oil derived from land-based crops [2]. With the need to reduce carbon emissions, and the decreasing reserves of crude oil, liquid fuels derived from plant material or what also known as biofuels, they are an attractive source of energy. Moreover, in comparison with other forms of renewable energy such as wind, tidal, and solar, liquid biofuels allow solar energy to be stored, and also to be used directly in existing engines and transport infrastructure. The term biofuel refers to liquid or gaseous fuels for the transport sector that are predominantly produced from biomass. Biofuels including bioethanol, biomethanol, biodiesel and biohydrogen appear to be attractive options for the future transport sector. Biodiesel is better than diesel fuel in terms of sulfur content, flash point, aromatic content and biodegradability [3]. Currently, bio-ethanol from, for example, corn starch, sugar cane or sugar beet, and biodiesel from oil crops such as palm and oilseed rape, are the most widely available forms of biofuel. However, there are two major issues over the sustainability of these first generation biofuels. The term biofuel refers to liquid or gaseous fuels for the transport sector that are predominantly produced from biomass. Biofuels including bioethanol, biomethanol, biodiesel and biohydrogen appear to be attractive options for the future transport sector. Biodiesel is better than diesel fuel in terms of sulfur content, flash point, aromatic content and biodegradability.

Biodiesel, defined as the monoalkyl esters of fatty acids derived from vegetable oil or animal fat, has demonstrated a number of promising characteristics in applications as an extender for combustion in compression–ignition engines [CIEs], including a reduction of exhaust emissions. In another definition, biodiesel is a diesel fuel derived from plant or animal oils, usually composed of methyl esters of long-chain fatty acids. The detailed chemical composition, in particular the chain length of the fatty acids, depends on the source of the oil. Biodiesel is usually produced from the oil by chemical transesterification, where the glycerol to which long-chain fatty acids are esterified in the source oil is replaced by another alcohol. This is usually, but not exclusively, methanol. Biodiesel can be used in standard diesel engines, but is often blended with conventional diesel.

Biodiesel is much less polluting than petroleum diesel, resulting in much lower emissions of almost every pollutant: carbon dioxide, sulfur oxide, particulates, carbon monoxide, air toxics and unburned hydrocarbons. Biodiesel does, however, have nitrogen oxide emissions that are about 10 percent higher than petroleum diesel. Blending biodiesel into petroleum diesel can help reduce emissions. It is well known that transport is almost totally dependent on fossil fuels. Biodiesel is one of the feasible alternatives. The biodiesel fuels have not been widely accepted in the market because they are more expensive than petroleum fuels. With recent increases in petroleum prices and uncertainties concerning petroleum availability, there is renewed interest in biodiesel fuels for diesel engines. Biodiesel is reported to be highly biodegradable in freshwater as well as soil environments. 90–98% of biodiesel is mineralized in 21–28 days under aerobic as well as anaerobic conditions. Biodiesel has been reported to remove twice the amount of crude oil from sand as conventional shoreline cleaners. Biodiesel increases the biodegradability of crude oil by means of co-metabolism. More than 98% degradation of pure biodiesel after 28 days is reported by some researchers in comparison to 50% and 56% by diesel fuel and gasoline respectively. Also, the time taken to reach 50% biodegradation reduced from 28 to 22 days in 5% biodiesel mixture and from 28 to 16 days in case of 20% biodiesel mixture at room temperature. The biodegradability of the mixture was reported to increase with addition of biodiesel. One method to produce the biodiesel is by extracting it from the microalgae and this way as we will see from many research papers proved to be a good and efficient way [4]. The advantages of algae as a source of biofuel may be summarized as follows: rapid growth rate compared to other sources, it grows practically anywhere, a high per-acre yield (7-31 times greater than the next best crop – palmoil), can be harvested daily for some species, it contains no sulfur and no toxic matter, the algae extracts can be used as livestock feed and even processed into ethanol, and may reduce carbon emission based on where it is grown. The main disadvantage is that it is considered as relatively new technology.

Energy Produced from Algae

While the general idea of using algae for energy production has been around for over 50 years, the concept of using lipids derived from algal cells to produce liquid fuels arose more recently. The research of liquid fuel produced from microalgae was begun at middle 1980s in 20 centuries. Aquatic biomass may represent a convenient solution, because it has a higher growth-rate than terrestrial plants. Microalgae have been extensively studied so far, as they can grow both in fresh- and salty-waters.

Microalgae can potentially be employed for the production of biofuels in an economically effective and environmentally sustainable manner. Algae can be used to produce biofuel, called algae fuel, algal fuel or even third generation biofuel. Compared with second generation biofuels, algal fuels have a higher yield: they can produce 30–100 times more energy per hectare compared to terrestrial crops. Microalgae have been investigated for the production of a number of different biofuels including bioethanol, vegetable oils, biodiesel, bio-oil, bio-syngas, and bio-hydrogen. The production of these biofuels can be coupled with flue gas CO_2 mitigation, wastewater treatment, and the production of high-value chemicals. Developments in microalgal cultivation and downstream processing are expected to further enhance the cost effectiveness of the biofuel from microalgae strategy. The high growth rate of microalgae makes it possible to satisfy the massive demand on biofuels using limited land resources without causing potential biomass deficit. Microalgal cultivation consumes less water than land crops. The tolerance of microalgae to high CO_2 content in gas streams allows high-efficiency CO_2 mitigation. Microalgal farming could be potentially more cost effective than conventional farming. Nitrous oxide release could be minimized when microalgae are used for biofuel production [5]. On the other hand, one of the major disadvantages of microalgae for biofuel production is the low biomass concentration in the microalgal culture due to the limit of light penetration, which in combination with the small size of algal cells makes the harvest of algal biomasses relatively costly. The higher capital costs of and the rather intensive care required by a microalgal farming facility compared to a conventional agricultural farm is another factor that impedes the commercial implementation of the biofuels from microalgae strategy [6].

Algae

Algae are organisms found in aquatic environment that are able to do photosynthesis. Usually they are eukaryotic organisms. The algae had evolutionary classes so they are very enormous in number and they are very diverse. Many research papers were dedicated for the purpose of seeking the most appropriate method for getting the maximum amount possible of the biodeisel from the microalgea. Here some points for optimizing the algal growth by controlling some of the factors needed for the algal growth [7].

Optimization of Algal Growth

Producing microalgal biomass is generally more expensive than growing crops. Photosynthetic growth requires light, carbon dioxide, water and inorganic salts. Temperature must remain generally within 293–303 K. To minimize expense, biodiesel production must rely on freely available sunlight, despite daily and seasonal variations in light levels.

The Importance of Light

The algal biomass is a distinctive indicator for how percentage benefit we got from the algae. Because the algae are photo synthetically active species that depend mostly on light to get their energy, that's why the light is the first factor that one can think of in changing and modulating the algal growth. If we are going to talk in terms of light energy and how much is used by the algae then for example: Eight photons of photo synthetically active radiation or about 48% of the incident solar flux are required to fix one molecule of CO_2 into carbohydrate, resulting in a maximum photosynthetic efficiency, that is not including respiration, of about 12%. When respiration is taken into account, the maximum efficiency falls to about 9%. Over short durations and in favorable conditions, low to moderate light levels, reported efficiencies of about 4.5–7% are typical in both ponds and photo bioreactors, which, in a pond, translate to a yield of 30–40 g of dry biomass per square meter of pond per day. In more simple words more light energy gives higher biomass which means more oil content and hopefully more biodeisel yield and the highest photosynthetic efficiencies are realized at low light intensities [8].

The control of light in the algal growth system might be seen as a hard job but it is actually can be done by controlling the living environment which might be either in the ponds or in photo bioreactors. Photo bioreactors, in contrast to ponds, offer the opportunity to optimize the light path, the extent to which the incoming light is diluted, and also the frequency of the light and dark cycle seen by an algal cell as it travels from deep in the culture to the illuminated surface. For example, using arrays of vertical annular columns, the productivity was significantly greater than using a horizontal pond.

Finally, enhanced efficiency of light utilization could result from the use of mixed cultures able to use different wavelengths of incident light.

Nutrient Supply

Several recent studies have described growth conditions that can be used for the industrial scale-up of lipid production from algae. The two-stage process suggested achieved 0.2 $kg/m^3/d$ oil for photosynthetic microalgae. In this set-up, cells were first grown under nutrient-sufficient conditions for biomass accumulation followed by nutrient deprivation for lipid synthesis. However, another study found that the most effective strategy for achieving high lipid contents in C. vulgaris was to allow cells to deplete their Nitrogen naturally rather than to transfer them to a medium completely lacking Nitrogen.

Many algae can grow heterotrophically with an external carbon source, rather than photosynthetically [9]. This alters the Nitrogen:Carbon ratio, and thus has the same effect as growing algae with reduced Nitrogen levels under autotrophic conditions, namely elevated lipid production. For example, Chlorella protothecoides had 55% of their weight lipids when grown heterotrophically, four times higher than under autotrophic conditions. A further enhancement was seen when sugars were added to photo synthetically grown cells of C. protothecoides, resulting in lipid levels of up to 11.8 $kg/m^3/d$. Similar to the protocols described above algae were grown in two phases: at the end of the log-phase, algal cells grown autotrophically were transferred to a medium containing 45 g/l glucose. At the same time, Nitrogen levels were reduced in order to induce lipid biosynthesis. Under these conditions, 58.4% oil content per dry weight was achieved, with no alteration in fatty acid composition. On the other hand, the source of sugars or other organic carbon is an issue, as is the potential for contamination of cultures by heterotrophic fungi and bacteria.

Maximizing the TAG Content in Algae

The yield of biodiesel from algae depends not just on the concentration of biomass achieved, but also on the oil content of the individual cells. In general, productivity and lipid content are inversely correlated, and stress conditions such as deprivation of Nitrogen or phosphate, which limit cell growth, also increase lipid content. The underlying principle is that where there is insufficient Nitrogen for protein production necessary for growth, excess carbon from photosynthesis is channeled into storage molecules, such as Triglycerides or also known as [TAGs] or starch and protein content may be reduced.

Metabolic Engineering of Lipid Production Pathways

With the advent of genome sequences and molecular tools for algae, there is the possibility that metabolic engineering may provide important and significant improvements for algal biodiesel production, for instance by increasing yields of TAGs, or engineering pathways for novel biofuel molecules. At present, we are far from understanding fully the detailed molecular biology and regulation of lipid body metabolism in algae, and although bioinformatics of sequenced algal genomes indicates that the same pathways are likely to operate, there has been little experimental verification of putative enzyme activities. Nonetheless, two recent papers have made some significant observations. Lipid accumulation can be induced by Nitrogen deprivation. This inducible process provides a useful experi mental basis for observing changes in gene transcript, protein and metabolic activities during lipid accumulation in algae. In cultures transferred to Nitrogen-depleted media, they were able to confirm that after 72 hours the total fatty acid content on a per cell basis increased by 2.4-fold, of which 65% of the total fatty acids were esterified to TAG in oil bodies, when compared to the initial culture. This indicated that TAG formation is increased by de novo synthesis of fatty acids or in other word by new synthesis of the fatty acids. Proteomic analysis, or the analysis of protein structures and function, identified a 'major lipid droplet protein' also known as [MLDP], which was highly abundant in the lipid bodies. The MLDP mRNA transcript abundance also followed the corresponding increase in lipid droplets after Nitrofen depletion. RNAi lines of C. reinhardtii with a 55–60% reduction of the MLDP transcript produced lipid droplets that had a diameter 40% larger than the control lines, implying that the function of this protein is to regulate lipid droplet size.

Another way of increasing lipid yield is to delete other 'redundant' or repeated pathways in the organism, so freeing precursor metabolites for the desired biofuel production. A paper have successfully measured an increase in the abundance of TAG lipid droplets in the C. reinhardtii starchless mutant , deficient in ADP-glucose pyrophosphorylase which is an essential enzyme in starch production. After 48 h of Nitrogen depletion, wild-type cells had increased their lipid droplet content by 15-fold, but there was a 30-fold increase in lipid droplets in the "sta6" starchless mutant algae. Ultimately, this resulted in a higher concentration of TAG per cell, as after 18 h of Nitrogen starvation the sta6 had on average 17 ng TAG [10^3 cells^{-1}] compared to 10 ng [10^3 cells^{-1}] in the wild-type cells.

The Biodiesel Production Cost

Microalgae production in closed photo bioreactors also known as [PBRs] is highly expensive. Closed systems are much more expensive than ponds. However, the closed systems require much less light and agricultural land to grow the algae. High oil species of microalgae cultured in growth optimized conditions of PBRs have the potential to yield 19,000–57,000 of micro algal oil per Acre per year. The yield of oil from algae is over 200 times the yield from the best-performing plant/vegetable oils [10].

Water, containing the essential salts and minerals for growth, is the second requirement. Fresh water is a valuable resource as are the salts and minerals needed; however, algae cultivation can be coupled to another type of environmental remediation that will enhance productivity while mitigating pollution. High nutrient wastewater from domestic or industrial sources, which may already contain nitrogen and phosphate salts, can be added to the algal growth media directly. This allows for algae production to be improved cheaply, while simultaneously treating wastewater. Alternately, salt water can be used, either from saline aquifer or sea water. This means that competition for water will be low.

Nutrients such as phosphorus must be supplied in significant excess because the phosphates added complex with metal ions, therefore, not all the added Phosphate is bio-available. Sea water supplemented with commercial nitrate and phosphate fertilizers and a few other micronutrients is commonly used for growing marine microalgae. Genetic and metabolic engineering are likely to have the greatest impact on improving the economics of production of micro algal diesel. Growth media are generally inexpensive. Microalgae biomass contains approximately 50% carbon by dry weight. All of this carbon is typically derived from carbon dioxide. Producing 100 ton of algal biomass fixes roughly 183 ton of carbon dioxide. Feeding controlled in response to signals from pH sensors minimizes loss of carbon dioxide and pH variations. Algae can grow practically anywhere where there is enough sunshine. Some algae can grow in Microalgae are very efficient solar energy converters and they can produce a great variety of metabolites. The culture of algae can yield 30 50% oil. Oil supply is based on the theoretical claims that 47,000–308,000 l/hectare/year of oil could be produced using algae. The calculated cost per barrel would be only $20. Currently, a barrel of oil in the U.S. Market is selling for over $100 per barrel. Despite all the claims and research dating from the early 1970s to date, none of the projected algae and oil yields has been achieved. Algae, like all plants, require large quantities of nitrogen fertilizer and water, plus significant fossil energy inputs for the functioning system.

Open ponds can be categorized into natural waters [lakes, lagoons, ponds] and artificial ponds or containers. The most commonly used systems include shallow big ponds, tanks, circular ponds and raceway ponds. The polymers can be used in very small amounts, without contributing a major cost to the overall process. The base case [30 g/m2/d] capital costs were estimated at almost $72,000/ha, without working capital, or almost twice as high as the prior effort. This was due to higher costs for many components, such as earthworks, which were several-fold higher. Among other things, higher costs were assumed for rough and fine grading, which depends on the type of site assumed to be available. Also the 1987 study estimated about $5000/ha to provide a 3–5 cm crushed rock layer, specified to reduce the suspension of silt from the pond bottom. There is, however, little evidence for a need for such erosion prevention, except perhaps for some areas around the paddle wheel and perhaps the turns. Further, there was a study that selected slip form poured concrete walls and dividers [baffles] as the design choice. A power generation system can be specified to produce electricity from the methane generated from the algal residues at about 10% of total costs. Finally we can conclude that even if the cost of the production of biodiesel from the microalgae was high, it would not be higher than the cost of the production of the oil itself. But the environmental benefits of using biodiesel worth the expenditure.

Scope of Work

The objective is the optimization of the growth of the microalgae in order to get maximum yield of the oil and thus the biodiesel form it. The light was the first and most important factor affecting the growth of microalgae. The nutrient and the amount of mineral found in the water or the feeding environment for the microalgae was the next important factor. The higher the content of the triglycerides [TAGs] the higher biodiesel production one can get from the microalgae. Finally I included the cost of modulating the factors to optimize the growth of microalgae. From the papers we can conclude that the production of biodiesel from the microalgae can be improved to very higher standards by controlling the factors that the growth depends on. The conclusion of other previous works is that the yield of biodiesel from algae depends on optimizing the growth of the algae. So it is important to optimize the growth process of algae and try different types of feed.

Project Objectives

- Design a prototype for growing algae with carbonated materials (out of power plant or waste water treatment unit) as a source for their food.
- Extract oil form algae.
- Convert raw oil to biodiesel.
- Compare properties of biodiesel to those of diesel.
- Test a diesel engine on the algae oil methyl ester (or raw algae oil).

Table 1. Different types of microalgae strains

Strain	Oil %	Medium
Chlorella minutissima	44	Soil/water
Tetraselmis	45.7	Soil/water
Neochloris oleoabundans	35-54	Nitrogen-limited medium
Chlorococcum pinguidum	35-54	Nitrogen-limited medium
Ankistrodesmus braunii	18.6-73	Nitrogen-limited medium
Chlamidomonos reinhardi	21	Soil/water

Choosing the Best Type of Microalgae

Table 1 illustrates the species currently being studied for their mass oil producing.

How to Choose the Type of Microalgae?

The type of microalgae could be selected depending on the strain properties mentioned below:

1. Toxic or nontoxic.
2. Percentage of oil per dry weight.
3. Culture or medium to grow in.
4. Availability.
5. Amount of nutrients necessary for growth.
6. Amount of sunlight necessary.

Where to Buy Algal Strain?

There are many places that sell algal strains however the best place is the University of Texas. The University of Texas is maintaining a large algal deposit located online at http://www.utex.org/. The culture collection includes approximately 3000 different strains of living algae. The primary function of UTEX is to provide algal cultures at good prices. The most important advantage from buying from the UTEX are easy to buy and can be sure of quality.

Final Selection of Microalgae

As mentioned above there is no strain of algae or at least one can be agreed on. Ankistrodesmus braunii is the best algal strain .It has a high oil content, good price and easy to cultivate.

Description for Ankistrodesmus Braunii Microalgae

Ankistrodesmus braunii is a unicellular green alga. Cells are solitary or loosely clustered in bundles or in tufts. Pure cultures of Ankistrodesmus braunii showed continual morphological fluctuations. They are needle-like; crescent-shaped, or narrowly tapering toward each end, sometimes straight, usually curved. The chloroplast is parietal, and has or does not have pyrenoid. Varying results were obtained in the determination of lipid content of Ankistrodesmus braunii in stationary phases, depending on the age of the culture and method of analysis. Total lipids ranged from 18.6 to 73 percent for cultures from 4 to 7 weeks of age respectively. Of the total lipids, 85.5 percent were recovered in the following distribution: nonsaponifiable, 17.1 percent; saponifiable, 61.5 percent; and glycerol, 6.8 percent.[1]

[1] UTEX The Culture Collection Of algae, (http://www.sbs.utexas.edu/utex/Search.aspx).

Cultivation of Microalgae Cultivation Process

It is fundamentally important to maintain stock of algae in order to investigate the optimal running conditions. The stock will be maintained in ideal culture and will be utilized to feed the different experiments with algae. A closed system which is called "photobioreactor" would be the most sufficient way to cultivate the microalgae stock. Photobioreactors enable controlling medias' temperature, pH number and the amount of CO_2. It also consists of a smart lighting system; therefore it will lead for more efficient cultivation process and more efficiency comparing to open systems.

Objectives

The main objective in designing the cultivation reactor is to cultivate the algae strain in suitable environment, and maintain the algae in the photobioreactor as stock for the next processes. To achieve this objective, it is very important to work on the direct requirements for the algae to reach the maximum growth rate. Designing the cultivation reactor started by putting all objectives on the paper. Objectives are what our desires from this design are. All objectives classified into 5 sections which are Growing the Algae, requirements, controlling, maintenance and economical. Table 2 shows the objectives table is shown below:

Table 2. Cultivation Reactor Objectives

Cultivation Reactor		
a) Growing Algae	b) Requirements	c) Controlling
d) Maintenance	e) Economical	
Growing Algae		
Size: having enough area for the algae to grow	Growing efficiently	Able to cultivate different type of microalgae
Requirements		
Easy to construct	Easy to feed with CO_2	East to feed with nutrients
Cultivation Reactor		
Smart lighting system	Movable	Easy to collect Algae
Stable	Runs on 220v	power supply
Maintenance		
Easy to maintain the reactor without damaging the Algae	Availability of spare parts	No leakage during the maintenance
Maintenance		
quality maintenance		
Controlling		
No leakage	No pollution	Controlling the amount of algae added or removed from the reactor
Controlling the temperature of the media	Controlling the pH	Controlling the amount of light supported to the media
Controlling the nutrients	Ability to stop the cultivation	Alarm system for the temperature
Digital Temperature measurements	Digital pH measurements	
Economical		
Less cost for the reactor	Less cost for spare parts	Inexpensive materials
Less nutrient costs	Less lighting system cost	Less maintenance cost
Low light power		

MARKET ANALYSIS FOR THE CULTIVATION PHOTOBIOREACTOR

Regarding to the microalgae cultivation reactor, study had been done on the available designs in the markets. Designs will be mentioned with their features. The most important feature for the photobioreactor is that the capacity is high enough which can be used as a store for the cultivated microalgae to use it in some next stages like producing the algal oil by feeding the algae with CO_2. Also it can work for continues operation or run which will save the process from any failure.

Features:

1. High capacity of microalgae.
2. Medium cost.
3. High performance.
4. Continues operation.

The fermentation apparatus is very useful reactor for cultivating the microalgae because the reactor is designed with very good mixing technique which will keep the algae with the media in better concentration. Fermentation apparatus has normal lighting system using spot lights

Features:

1. Easy media exchange.
2. Rotating container vessel.
3. Enhanced with motor.
4. Enhanced with heater.

The advantage of the Box-type vessel is that it is big enough to give you the maximum capacity for algae cultivation. Also it is easy to use.

Features:

1. Simple materials.
2. Large capacity.
3. Easy to use.
4. Enhanced with cooling coil.

The spinning tube-in-tube photobioreactor is considered very efficient reactor because of the high technology techniques used in designing this photobioreactor. First it has smart lighting system with evacuated glass gaps to ensure the light to penetrate to the maximum distance inside the reactor; also it is designed with uncoupled mixing and aeration process to distribute the CO_2 and O_2 to every point in the reactor.

Features:

1. High technology used.
2. Smart lighting system.
3. Enhanced with motor.

4. Temperature Control.
5. Easy to use.
6. Uncoupled mixing and aeration.

Function Analysis

The functions of the cultivation reactor are shown in Table 3:

Table 3. Cultivation Reactors' Function Analysis

Receive water	Receive nutrients	Collect algae
Media exchange	Grow algae	Alarm system for the temperature.
Lighting system	Move the reactor	Receive CO_2/waste water
Testing the media	Remove the dead algae	Control the temperature
Control pH	Electricity supply to the reactor	Prevent the leakage
Measuring the temperature	Measuring the pH	

Selected design and description

An evaluation table has shown that the best design of cultivation reactor should have the features below:

1. Cylindrical design.
2. Controlled mixing and aeration.
3. High biomass density.
4. High productivity.
5. Built in baskets made of neon as smart lighting system.
6. Controlling the CO_2 inlet.
7. Controlling PH by the amount of nutrients.
8. Less cost of the raw materials.
9. Less cost for the maintenance.
10. Available spare parts.
11. Light intensity 50 μmol m^{-2} s^{-1}.
12. Rounded with spot lights with the same light intensity.
13. Inside coils for heating and cooling to the desired temperature.
14. The maximum cell density 0.5~1.2 * 10^6 cells ml^{-1}.

BIO REACTOR TO PRODUCE OIL FROM MICROALGAE (USING CO_2 AND WASTE WATER)

Introduction to the Bioreactor

Algae can also be grown in a photo bioreactor. A photo bioreactor is a bioreactor which incorporates some type of light source. However the term is more commonly used to define a closed system, as opposite to an open tank or pond. Photobioreactor design will enhance algal biomass production by keeping algae genetics pure and reducing the possibility of pollution. All essential nutrients must be introduced into the system to allow the microalgae to grow and to be cultivated to give the maximum amount of oil. Essential nutrients include carbon dioxide, nitrates, waste water, minerals and light. A photo bioreactor is also possible to introduce a continuous flow of waste water containing nutrients, and carbon dioxide. For a photo bioreactors, the better controlled environment will lead to more volumetric biomass and more oil efficiency.

OBJECTIVES

Designing the cultivation reactor started by putting all objectives on the paper, without any corrupting. Objectives are what our desires from this design are. All objectives classified into 5 sections which are producing oil, requirements, controlling, maintenance and economical. Table 4 shows the objectives of the cultivation reactor:

Table 4. Bioreactor Objectives

Objectives of Bioreactor for producing oil using CO_2/Waste water		
Increasing oil contents	Requirements	Controlling
Maintenance	Economical	
Increasing oil contents		
More oil contents	Easy method to provide CO_2	Deals with different types of microalgae
Requirements		
Easy to construct	Compact size	Easy to feed with CO2 or Waste water
No environmental effect	Movable	Easy to collect Algae baskets
Stable	Lighting system	
Maintenance		
Easy to maintain the bioreactor without damaging the Algae	Availability of spare parts	No leakage during the maintenance
Easy to maintain without complexity		
Controlling		
No leakage	No pollution	Controlling the amount of algae in the baskets
Monitoring the temperature of the media	Monitoring the pH	Controlling the amount of CO2
Ability to stop the bioreactor	Controlling the Waste water	Controlling the amount of light supported to the media
Economical		
Less lighting system cost	Inexpensive spare parts	Inexpensive materials
Low maintenance cost		

FUNCTION ANALYSIS OF BIOREACTOR

1. Getting maximum amount of oil.
2. Receive algae from cultivation reactor.

3. Collect the baskets.
4. Move the bioreactor.
5. Feed CO_2.
6. Remove the dead algae.
7. Electricity supply.
8. Prevent the leakage.
9. Start and stop.

OIL EXTRACTION

Oil extraction process aims to separate the oil from algae cells. It is a process which can determine the success of producing the biodiesel. Once the algae have been grown and its oil content reached the maximum amount, extraction process takes place. It depends on the success of cultivation process while increasing the algae cells will produce more oil. The bioreactor which increases algal oil content, effects on the oil extraction efficiency because it leads for more oil production. The cost, efficiency, toxicity, and the ease of oil extraction process are the most aspects to be considered in oil extraction design. Chemical solvent extraction will take place after the oil is extracted from algae cells because oil will remain mixed with algae cells. Therefore, chemical solvents should separate the oil from the mixture and its toxicity and cost should be considered in selecting the type of the solvent.

Extraction Methods

There are many methods to extract oil from algae and it should be chosen due to algae type and the cost of the project:

1. Expeller/Press.

When algae are dried it keep its oil content, which can be pressed out with a mechanical expeller. By this method oil can be extracted from algae in a simple process and it can extract 70-75% of the oil. It is an easy method of extracting oil but it is not efficient.

2. Chemical solvent extraction.

Oil can be extract from algae by using chemicals such as benzene, ether and hexane. Hexane is the most popular chemical used due to its low cost. The problem with it is due to the hazardous of chemicals but it is good method for small scale.

3. Supercritical Fluid extraction.

This method can extract mostly most of the oil. CO acts as the supercritical fluid which is pressurized and heated to have liquid and gas specification in the same time. When CO_2 and algae are combined, the CO_2 turns the algae completely into oil. This method is efficient but it needs special equipment and it is expensive.

4. Enzymatic extraction.

Enzymatic extraction uses enzymes to break down the cell walls with water acting as the solvent which makes fractionation of the oil much easier.

5. Ultrasonic-assisted extraction.

This method can greatly accelerate the process. By using an ultrasonic reactor the ultrasonic waves are created cavitations bubbles in a solvent material which collapse near the cell walls, it creates shock waves and liquid will break cells wall and release their oil contents into the solvent.

Table 5 below compares between all these methods with considering the cost, toxic, and the ease of the process.

Table 5. Extraction of oil methods

Method	Advantage	Disadvantage
Expeller/press	Easy Low cost	Not so efficient Do not get all oil from the algae cell
Chemical solvent	Good for small scale Extract most of the oil	Danger May cause unwanted reaction Waste chemical (Pollution) Further separation is needed
Supercritical fluid	Fast method Efficient No waste chemicals	Special equipment High cost
Enzymatic	Efficient	Needs enzymes grow Work with other method like chemical solvent Needs control condition Needs further separation High cost
Ultrasonic-assisted	High efficient Fast	Heat generation Requires cooling

SELECTED METHOD FOR EXTRACTION

The selected method is crushing (pressing) because it is easy to do and the cheapest method. Harvested algae will be crushed to produce the oil from its cells while hexane solvent will follow it to separate the oil from algae-oil mixture. The reason of choosing hexane solvent is that it will volatilize after separating the oil from algae cells while it is not a toxic solvent.

CATALYZATION

Catalyzation is a process of using the catalyst with the reactants to increase the rate of a reaction without being consumed in the process.

There are three popular methods the get biodiesel:

- Base catalyzed transesterification of the oil.
- Direct acid catalyzed transesterification of the oil.
- Conversion of the oil to its fatty acids and then to biodiesel.

The most useful method today is Base catalyzed transesterification of the oil due to it is low temperature and pressure, easy, direct conversion without intermediate compounds and to yields high conversion (98%) with minimal side reactions and reaction time.

CONVERTING THE ALGAE OIL TO BIODIESEL

After extracting the oil from algae, biodiesel will be produce from oil in a reaction called transesterification (Base catalyzed transesterification):

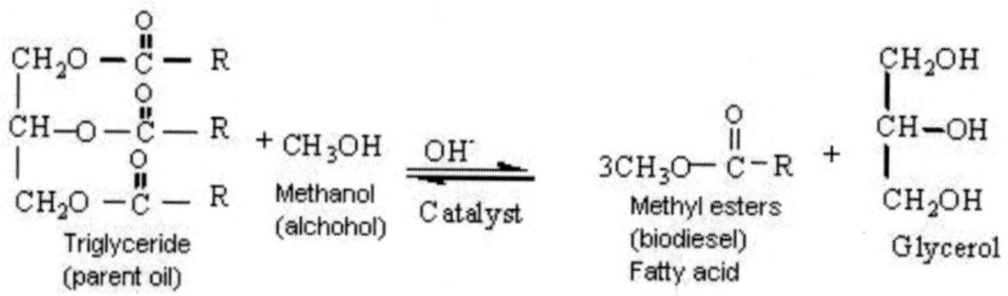

Transesterification is an economical process to get biodiesel from oil because it requires low temperature and pressure. The transesterification is the reaction of triglyceride with alcohol to form methyl ester (biodiesel) and glycerol.

Triglyceride reacts with methanol .The reaction done step wise. First, triglyceride converts to diglyceride then to monoglyceride. The reaction happens with catalyst presence. The catalyst usually used is sodium or potassium hydroxide which is base.

Table 6. Laboratories Testing Services for Biodiesel

Flash Point	Cetane Index	Sulfur Content, % Wt.
Water and Sediment	Distillation	Carbon Residue, % Wt.
Kinematic Viscosity	Cloud Point	Distillation, °C
Sulfated Ash	Acid Number	API Gravity
Sulfur Content	Free Glycerine	Carbon Residue
Copper Corrosion	Total Glycerine	Fuel Lubricity
Cetane Number	Phosphorous Content	Potassium and Sodium

The reaction is reversible so excess methanol should be added to form more biodiesel and to restrict any reverse reaction. The reaction will take 90 minutes to finish in 60°C in atmospheric pressure. The product will be a biodiesel and glycerol which do not mix. The biodiesel will be in the top while glycerol will be in the bottom due to their density. So, biodiesel and glycerol can easily separated from each other. Some of very little amount of soap will form due to the reaction between oil, water and catalyst and it will form a small layer of soap between biodiesel and Glycerol. To prevent saponification reaction which cause yield loss, oil and methanol should be dry.

In the beginning, knowing the molecular weight of the oil is not clear. So, a rough amount will be start on in the first run 12.5% of methanol to 87.5% of oil. Nothing will be loss because as reactants will be in the flask will give a product. The target is to get one liter of biodiesel as minimum amount.

TESTING THE PRODUCED BIODIESEL

After collecting the produced biodiesel form the conversion process, there are some steps for testing the produced oil.

Step 1:

Find all the international standards and regulations for the physical and chemical properties for the biofuel in general.

ASTM and European fuel standards:

1. ASTM D6751 Biodiesel Fuel Blend Stock, B100.
2. EU biodiesel fuel standard EN14214.
3. Biodiesel feedstock evaluation and testing to AOCS methods.

Step 2: Find the physical and chemical properties for the produced biodiesel in the laboratory. To check if the produced oil can be considered to be biofuel due the international standards. Some laboratories testing services for biodiesel are listed in Table 6.

Step 3: Run the produced biodiesel in a normal diesel engine in the lab, and measure the exhausted emissions and calculate the output power the efficiency. After that comparison should be conducted between the results from produced biodiesel and the results from the petroleum diesel.

Preparation of Algae Oil and Its Methyl Ester

Microalgae Used for Oil Extraction

Two types of microalgae were used for oil extraction namely Ankistrodesmus braunii and Nannochloropsis. Ankistrodesmus braunii was obtained from the Algae Collection Centre, The University of Texas at Austin. Nannochloropsis sp was obtained from a local research centre in Umm Al Quwain, UAE. Microalgae are obtained in small amounts and have to be cultured in culture medium to obtain sufficient cell counts before the microalgae could be transferred to larger bioreactors or open ponds. Ankistrodesmus braunii and Nannochloropsis sp. were used in two large-scale open ponds.

Culture Medium

The culture medium used for the primary growth of algae was Modified Bold 3N medium containing the following components: Sodium Nitrate ($NaNO_3$), Magnesium Sulphate 7-Hydrate ($MgSO_4.7H_2O$), Sodium Chloride (NaCl), Dipotassium Hydrogen Phosphate

(K_2HPO4), Potassium dihydrogen orthophosphate (KH_2PO_4), Calcium Chloride 2-Hydrate ($CaCl_2.2H_2O$), P-IV Metal solution (containing Na_2EDTA, $FeCl_3.6H_2O$, $MnCl_2.4H_2O$, $ZnCl_2.6H_2O$, $CoCl_2.6H_2O$, and $Na_2Mo_4.2H_2O$), Soil water, ***Vitamin B1 and Vitamin-B12***. The algae were first grown in 500mL conical flasks under fluorescent light system with aeration at regular intervals to obtain sufficient cell counts and then in 5L culture medium under the same culture conditions. After the cultures have sufficient microalgae cell counts, the algae are then transferred to the open pond system. Treated water from sewage plant is used in the open pond system and the feed provided is raw sewage water which provides the nitrogen, phosphates and other nutrients for the growth of microalgae.

Monitoring of Cell Count

The cell count is monitored on a daily basis to determine the growth of algae. On the basis of the cell count, the microalgae are transferred to a larger volume bioreactor. The cell count is monitored using Neubauer Counting Chamber. The cell density is calculated using the following formula:

$$d\,(\text{cells/ml}) = \frac{\text{total count}}{\text{No. of blocks}} \times 10^4$$

Open Pond System

When the algal cell count reached 10^6 cells/mL, the microalgae were transferred to the open pond system. The ponds have paddle wheels to ensure water circulation. 100L of raw sewage water was added to 15000L capacity pond as a source of nutrients. The cell count of the pond was monitored and when the cell count reached 10^9 cells/mL the algae was separated from the water and processed for oil extraction. The water and nutrients were replenished based on requirements.

Separation of Algae from Water

Three methods were used for separation of algae from water, they are: Centrifugation, Drying and Filtration. In the centrifugation process, 50mL tubes with the algae-water mixture were loaded in the centrifuge machine to obtain the sediment of algae due to centrifugal force. The algal sediment obtained were further dried to remove moisture and used for the oil extraction process.

In the drying process, containers (pans) of dimensions 1m x 1m x 10cm were used to dry the microalgae under the sun. In the filtration process, filters of 1micron pore size or less were placed in the pond to collect the algae and were further dried for oil extraction.

Oil Extraction

Two methods for oil extraction were employed: Ultrasonication and Soxhlet extraction. Sonication is the act of applying sound energy to agitate a sample, for various purposes. Sonication is often used to disrupt cell membranes and release cellular contents. Branson Sonifier 450 is the equipment used for sonication. Ultrasonication was used to rupture the cell wall of microalgae resulting in the release of algal oil. Ultrasonification is done in the presence of hexane so that the oil released dissolves in the hexane. After the sonication process the Hexane–Algae mixture is centrifuged to separate the algae from hexane. The hexane thus obtained has dissolved oil which has to be separated. Hexane has a lower boiling point than oil and hence can be evaporated at the boiling point of hexane leaving behind algal oil. In the Soxhlet extraction method hexane is evaporated and circulated through the powdered and dry algae sample placed in the Soxhlet apparatus. The vapors of hexane condense, mix with the algae dissolving the oil and finally return to the bottom where there is boiling hexane. The cycle is repeated several times. Finally the hexane-oil mixture is removed from the apparatus and hexane is evaporated off to give algal oil.

12-15% amount by volume of oil was extracted from Ankistrodemus braunii and 10% from Nannochloropis sp. using the two methods. The used biodiesel in the current work has been extracted from the Nannochloropis sp.

Table 7 illustrates the chemical composition of the algae-based raw oil as measured by high performance liquid chromatography. Similar to the oil-based feedstock, it generally consists of three main fatty acids: palmitic acid, oleic acid and linoleic acid. Most of the biofuels consist of these three acids. Palmitic acid is one of the most common saturated fatty acids found in animals and plants. As its name indicates, it is a major component of the oil from palm trees. Conversely, soybean oil and rapeseed oil consists of oleic acid and linoleic acid.

Transesterification of Algal Oil

The process of converting vegetable and plant oils into biodiesel fuel is called transesterification, and it refers to a reaction between an ester of one alcohol and a second alcohol to form an ester of the second alcohol and an alcohol from the original ester, as that of methyl acetate and ethyl alcohol to form ethyl acetate and methyl alcohol.

Chemically, transesterification means taking a triglyceride molecule or a complex fatty acid, neutralizing the free fatty acids (around 13% in the current raw oil), removing the glycerin and creating an alcohol ester. This is accomplished by mixing methanol with sodium hydroxide to make sodium methoxide. This liquid is then mixed into vegetable oil. The entire mixture then settles. Glycerin is left on the bottom and methyl esters, or biodiesel, is left on top. The methyl esters is washed and filtered.

The algal oil has been converted to Fatty Acid Methyl Esters (FAME) by the trancesterification process. The extracted algal oil is preheated to 60^0 C and added to a mixture of methanol and sodium hydroxide. It was then mixed with a blender for an hour. The amount of methanol used was 20% (ME0.2) of the amount of algal oil used. 3.5g of sodium hydroxide was used for a liter of oil. After the tranesterification process the mixture obtained is collected and kept for 8 hours or more to allow the glycerin to settle at the bottom and the Fatty Acid

Methyl Ester (Biodiesel) will form the top layer. The biodiesel is separated and washed with water 4 to 5 times and was heated or dried in air to evaporate the remaining water. It is important to remove most of the water in the oil as any remaining water may affect the combustion process and hence affects the engine operation. The water content was found to be less than 0.20% in the current methyl ester which is typical for biodiesels.

Another version of biodiesel has been also produced with the amount of methanol used is 10% (ME0.1) and the other amounts were the same. The physical characteristics of the algae oil methyl ester (ME0.2) may be seen in Table 8 with the corresponding measuring standard.

Table 7. Chemical Analysis of the Algae Raw Oil

Fatty acid methyl esters	Concentration (mg/100ml)
Myristic	49.8
Palmitic	6020
Stearic	6.1
Oleic	422.6
Linoleic	1900.2
Linolenic	17.7
Arachidic	6.9
Behenic	270.8

Table 8. Physical and chemical properties of algae oil methyl ester

Property	Value	Method
Initial boiling point, C	71.8	ASTM-DP86
Heating value, kJ/kg	40.72	ASTM DP 2015
Cloud point, C	7	ASTM-DP 2500
Pour point, C	-6	ASTM-DP97
Flash point, C (closed cup)	144	ASTM-DP 3828
Density, g/cm^3 (at 40°C)	0.869	ASTM-DP 1298
Dynamic viscosity, mPa.s (at 40°C)	3.64	ASTM-DP 445
Kinematic viscosity, mm^2/s (at 40°C)	4.19	ASTM-DP 445
Dynamic Viscosity (raw oil), mPa.s (at 40 °C)	48.5	ASTM-DP 445

Engine Experimental Apparatus

The research engine used in the present study is the research engine Ricardo E6 single cylinder variable compression indirect injection diesel engine.

The engine is loaded by an electrical dynamometer rated at 22 kW and 420 V. The engine is fully equipped for measurements of all performance parameters and combustion noise data. The combustion pressure time history is measured by a water-cooled piezo-electric pressure transducer connected to the relevant amplifier. The liquid fuel flow rate is measured digitally by a multi-function micro processor-based fuel system, Compuflow System. Two data acquisition systems are used to collect the important data and store it in two personal computers for offline analysis. The following parameters are fed into the first data acquisition system: liquid fuel flow rate, engine speed and torque, and air/oil/coolant/exhaust temperatures.

A computer program in μMACBASIC language is written to collect the data and manage the system and a workstation operating system has been used to run the program. Measurement of combustion pressure and crank angle are performed by another data acquisition system which is used to collect the cylinder combustion pressure and crank angle data.

The pressure signal is collected by a high pressure water cooled AVL piezo-electric pressure transducer and its output is fed into a charge amplifier then to a data acquisition card linked to the personal computer. The crank angle signal is fed into a degree marker shaper channel and the output is fed into the acquisition card. The acquisition card could collect data at the rate of 250 kHz. A LabVIEW program has been written to collect the data from the two channels at a sampling rate of 10,000 points per second and store the pressure and crank angle data in the computer disk for offline analysis.

A computer program is written in LabVIEW to find the maximum combustion pressure and the pressure rise rate data at all cycle points from mid compression stroke to mid expansion stroke. The maximum value of pressure rise rate is then obtained and recorded. This value will be used to represent the noise level at that operating condition. The Experiments have been carried out after running the engine for some time until it reaches steady state and oil temperature is at 55° C ±5, and cooling water temperature is at 65° C ±5.

Experimental Procedure

The study is carried out to cover five different types of fuels: base diesel fuel, algae oil methyl ester (0.2), algae oil methyl ester blended with diesel at 50/50 ratio, raw algae oil, and algae oil methyl ester (0.1). For each type of fuel the engine parameters are varied according to the following listed levels:

- The engine speed and it is varied from 18 to 30 rev/sec, incremented by 3 rev/sec
- The engine injection timing for the pilot fuel and it is varied from 20 to 45° BTDC, incremented by 5^{0}
- The engine load output torque and is varied from 2 to 18 Nm.

The data presented are maximum pressure rise rate, maximum pressure, engine output torque, maximum heat release rate and its crank angle of occurrence for the specified testing conditions.

The maximum uncertainties in the measured quantities were 2% in the engine speed, 4% in the engine torque, 6% in the mass of fuel, 1° in the fuel injection angle, 0.01 in the engine compression ratio, 2.5% in the maximum pressure and 3% in the pressure rise rate.

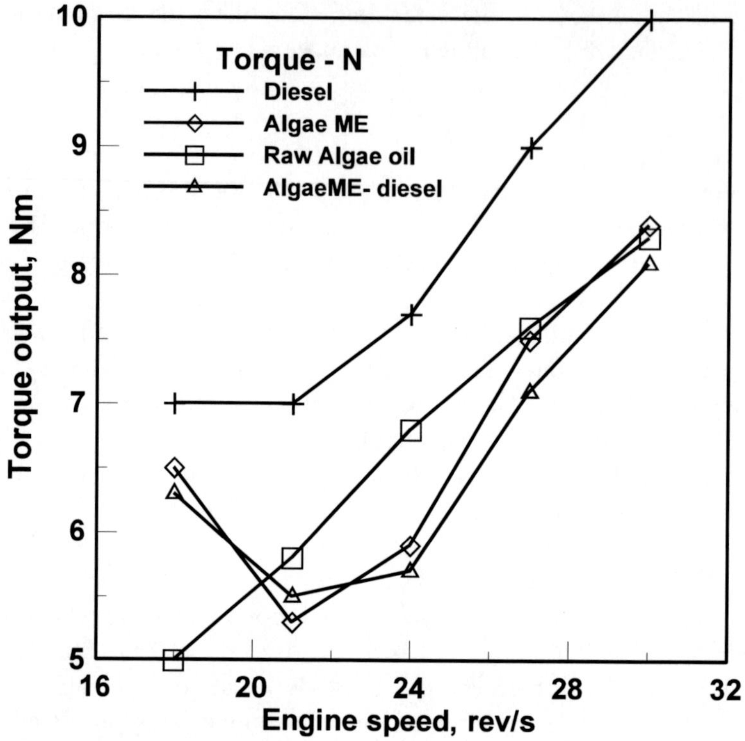

Figure 1. Effect of engine speed on torque output.

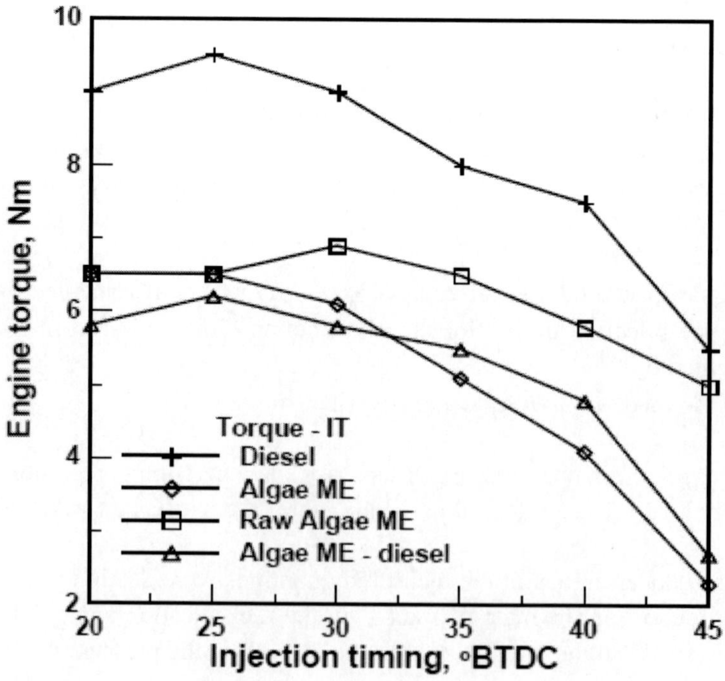

Figure 2. Effect of fuel injection timing on output torque from the engine.

RESULTS

Figure 1 illustrates the effect of engine speed on the engine output torque. Figure 2 shows the effect of fuel injection timing on the torque output of the engine. One common point which is easy to grasp is the reduction of the engine torque when the algae oil methyl ester is used. The highest torque output was for the diesel fuel. For raw algae oil, algae oil methyl ester and for algae oil methyl ester diesel blend of 50/50%, the output torque was lower. One possible reason for less torque output – at the same engine other conditions – is the delayed ignition of the fuel which leads to self-ignition of more mass of injected fuel. The increased mass of liquid fuel when burnt spontaneously causes the combustion pressure to increase sharply at a higher rate during the compression stroke of the engine. If the maximum pressure occurs during the compression stroke, it causes more resistance to the piston motion and increases the negative work and the resulting net output torque will reduce. The output torque /power of the engine may be improved by changing the engine parameters e.g. the compression ratio and injection timing as shown in figure 2.

CONCLUSION

From the experimental study conducted in the present work, the following conclusions may be withdrawn:

1. Algae selected proved to be a good source of biodiesel for diesel engines.
2. Three types of Algae have been selected and they produced acceptable yield of raw oil biofuel has been produced.
3. The physical and chemical properties of the produced Algae oil biofuel are close to the required biodiesel fuel.
4. The algae oil ME produced less engine torque output than the diesel case and raw algae oil.
5. Using less methanol (10%) in producing algae oil methyl ester resulted in a better performance of the engine.
6. The decrease in output torque can be improved by modifying the diesel engine conditions for the algae oil cases.

REFERENCES

[1] Fedorov, A.S., S. Kosourov, M.L. Ghirardi and M. Seibert, 2005. Continuous H2 photoproduction by Chlamydomonas reinhardtii using a novel twostage, sulfate-limited chemostat system. *Appl. Biochem. Biotechnol.,* 124: 403-12.

[2] Sarmidi Amin, Review on biofuel oil and gas production processes from microalgae, *Energy Conversion and Management,* Volume 50, Issue 7, Pages 1834-1840, July 2009.

[3] A.E. Atabani, A.S. Silitonga, Irfan Anjum Badruddin, T.M.I. Mahlia, H.H. Masjuki, S. Mekhilef, A comprehensive review on biodiesel as an alternative energy resource and

its characteristics, *Renewable and Sustainable Energy Reviews,* Volume 16, Issue 4, Pages 2070-2093, May 2012.

[4] Y. Haik, Mohamed Y. E. Selim and T. Abdelrahman, Combustion of algae oil methyl ester in an indirect injection diesel engine, Energy, *The International Journal*, Volume 36, Issue 3, pp. 1827-1835, March 2011.

[5] I. Rawat, R. Ranjith Kumar, T. Mutanda, F. Bux, Dual role of microalgae: Phycoremediation of domestic wastewater and biomass production for sustainable biofuels production, *Applied Energy,* Volume 88, Issue 10, Pages 3411-3424, October 2011.

[6] Man Kee Lam, Keat Teong Lee, Microalgae biofuels: A critical review of issues, problems and the way forward, *Biotechnology Advances,* Volume 30, Issue 3, Pages 673-690, May–June 2012.

[7] Esra Sorguven, Mustafa Özilgen, Thermodynamic assessment of algal biodiesel utilization, *Renewable Energy,* Volume 35, Issue 9, Pages 1956-1966, September 2010.

[8] Yingfu Guan, Maicun Deng, Xingju Yu, Wei Zhang, Two-stage photo-biological production of hydrogen by marine green alga Platymonas subcordiformis, *Biochemical Engineering Journal,* Volume 19, Issue 1, Pages 69-73, 1 July 2004.

[9] Xian-Ming Shi, Hui-Jun Liu, Xue-Wu Zhang, Feng Chen, Production of biomass and lutein by Chlorella protothecoides at various glucose concentrations in heterotrophic cultures, *Process Biochemistry,* Volume 34, Issue 4, Pages 341-347, June 1999.

[10] Ayhan Demirbas, M. Fatih Demirbas, Importance of algae oil as a source of biodiesel, *Energy Conversion and Management,* Volume 52, Issue 1, Pages 163-170, January 2011.

In: Algae
Editors: Dagmar Krueger and Helga Meyer

ISBN: 978-1-62081-280-8

Chapter 4

TRACHYDISCUS MINUTUS: A NEW ALGAL EPA PRODUCER

Jaromír Lukavský*
Acad. Sci. Czech Rep., Inst. Botany, Dep. Plant Ecology,
Biorefinery Research Centre of Competence, Třeboň, Czech Republic

ABSTRACT

The alga *Trachydiscus minutus* (Bourr.)Ettl, strain Lukavský et Přibyl 2005/1, Eustigmatophyceae, is described here as a new source of polyunsaturated fatty acids (PUFAs), especially of eicosapentaenoic acid (EPA). These fatty acids (FA) are valuable food additives for man, animals and in aquaculture.

T. minutus was first described as *Pseudostaurastrum minutum* by Bourrelly (1951), from a small fishpond in France, in 1951. It is now found in various water localities, e.g. ponds, lakes and streams, but usually as a rare species. Its principal mode of reproduction is by autosporogenesis, and in prolonged darkness, zoospores are produced. Only chlorophyll_*a* was evaluated; other pigments are violaxanthin, vaucheriaxanthin ester, and ß–carotene.

In laboratory cultures, biomass can reach 16 $g.l^{-1}$ dry weight (DW), consisting of 26-31% total lipids (TL) and EPA comprises up to 42% of TL. The TL content can be increased to 39% of DW by nutrient starvation (N,P,S deficiencies) resulting in an EPA content of up to 35.8% DW, and a productivity of 88 mg $l^{-1}.d^{-1}$. Targeted stereoisomers of triacyglycerols (TAGs) can also be obtained. Oil drops are prominent and an iodine value of about 216 was determined by Raman microspectroscopy, even in individual cells; this has high potential value for the food industry. *T. minutus* also contains 26% of saturated myristic fatty acid, and about 48% protein suitable for food and fuel. The optimal growth temperature was about 27 °C, but it survived below 15 and at 35 °C. Extreme temperatures resulted in increased TL content, up to 35% DW, and decreased carbohydrate (13.6% DW) and protein (30-38% DW). High temperature elevated the activity of superoxide dismutase, catalase and proteases, but glutathione reductase, and peroxidase activity decreased. Under 15 °C, all enzymes except glutathione reductase were inhibited. In the laboratory, optimal growth occurred under continuous light of 2x132 μmol $photons.m^{-2}.s^{-1}$, whereas the maximum light levels evaluated in the greenhouse were up to 640 μmol $photons.m^{-2}.s^{-1}$.

In an orbicular cultivation unit inside a greenhouse in N.Greece (vol. 200 litre, area 2 m^2) *T. minutus* produced 0.6 $mg.m^{-2}$/day i.e. 3.4 $mg.l^{-1}.day^{-1}$ of EPA. In SW. Bohemia, using a thin-layer inclined photobioreactor (volume 150 litre, area $15m^2$) in a greenhouse, *T. minutus* produced up to 0.25-0.45 $mg.m^{-2}.day^{-1}$, i.e. 20-30 $mg.l^{-1}.day^{-1}$ of EPA. *T. minutus* proved to be tolerant of manipulation, e.g. centrifugation, oscillations of light, temperature and CO_2 concentration, contamination with another algae etc. Sedimentation followed by centrifugation and lyophylization, or drying in the sun on plastic foil was suitable for harvesting. Flocculation also has potential (Přibyl pers. com.) and no parasites have yet been observed in *T. minutus.*

T. minutus was successfully immobilized in agar or alginate. Its potential for further research and exploitation is high and comparable with, or better than other algae used for biotechnological production of PUFAs, and especially EPA.

INTRODUCTION

Domesticated cyanobacteria and microalgae with potentially highly valuable properties, remain limited to a few genera. Primarily, production has focussed on mass cultivation of low cost, protein-rich food, but this has been expanded to include the production of nutrient additives and valuable pharmacological compounds e.g. proteins and pigments, phycocyanin from *Arthrospira* (*Spirulina*), chlorophylls from *Chlorella*, astaxanthin from *Haematococcus*, or carotenes from *Dunaliella* (Borowitzka and Borowitzka, 1988). Reviews are also in Raja (2008), and Richmond (1984, 2004).

The aim of this paper is to review studies on the growth and composition of *Trachydiscus minutus,* a new and potentially valuable species with an emphasis on the biotechnological production of PUFAs, especially eicosapentaenoic acid (EPA).

OLEAGINOUS ALGAE

Special attention has been paid to algae producing oils. The first experiments were carried out during World War II using Bacillariophyceae, with the aim of producing lubricants (Ketchum & Redfield 1938, Harder & Witsch; 1942a,b; Řetovský, 1946). The last author established outdoor cultures of *Navicula* sp. in a 70 litre closed glass cylinder.

The importance of high quality oils for the human diet, especially PUFAs, is well known (Borowitzka, 1988) but PUFAs are also important in aquaculture as fish feed via zooplankton (Borowitzka (1997), Brown (2002), Krienitz & Wirth (2006)). The search for new prospective producers of healthy oils, especially those rich in PUFAs, remains therefore a focus in phycology.

BIODIESEL FROM ALGAE

Prospects for the exploitation of algae as biofuel oscillates along with oil costs, international relations, politics of self-reliance and issues of national security (Mata et al. 2010). The alga with the highest potential, *Nannochloropsis* sp. from Eustigmatophyceae, was tested by Rodolfi et al. (2009) and *Scenedesmus obliquus* was proposed by Mandal &

Malik (2009). However, the existence of enormous oil supplies in oil-rich sands and bitumen slates has raised the potential for long-term mining prospects (Lomborg 2006). Programmes on algal biodiesel (Sheehan et al. 1998) have therefore been frozen at present (Pienkos & Darzins 2009).

Nevertheless, a price reduction of algal oil can be expected through combinations of novel waste water treatments e.g. *Botryococcus* (Li et al. 2008; Shen et al. 2008). CO_2 can also be supplied cheaply (free of charge) as flue gas from municipal waste incinerators, (Douskova et al. 2008).

The advantage of this approach is that it avoids payment of penalties for water pollution by wastes, or air by CO_2, and in parallel, produces valuable compounds. All possibilities hold prospects for industry, and provide feed for animals although this may be problematic as a food supplement for people, where not only the composition but also the origin of food is strictly controlled. CO_2 is also available free of charge from mineral springs or thermal bores, and providing hot water for heating the suspension culture (Furnadzhieva et al. 2002a,b). Reviews are presented by Courchesne et al. (2010), Huang et al. (2010), Narendra et al. (2010), and by Shi et al.(2011). Unfortunately, Petkov (2012) has calculated that the total energy spent in the production of algal biodiesel is much greater than the energy harvested at the end. Only under certain circumstances can production of algal biodiesel be economically viable.

Some algae, such as *Chlorella protothecoides,* can be cultivated heterotrophically e.g. in sugar cane juice, and the oil content can reach 53% DW (Cheng et al. 2009). The advantage of heterotrophic cultivation is that it is independent of climate and solar irradiation. Other improvements and further progress can be expected from the application of transgenics (Gressel 2008, Leon-Banares et al. 2004).

Trachydiscus minutus oil is not appropriate for algal biodiesel production because of its composition of mainly unsaturated fatty acids e.g. EPA (iodine value = 216), but it has potential as a food supplement. Meanwhile, its content of saturated myristic acid (26-28% of total lipids) also makes this alga a candidate for the cosmetic industry or for esterification into fuel (Table 2).

Algae as Feed in Aquaculture

An artificial nutrition chain: alga > zooplankton > fish was tested using a few algal species. *Nannochloris* sp., a marine chlorophyte alga was examined. This contains 7% DW of total oils, up to 23% of which comprise arachidonic acid and EPA (Petkov et al. 1994). The main sterols were cholesterol and sitosterol, and these, as well as fatty acids, were stable if $NaNO_3$, KNO_3, or urea were used as N sources. *Nannochloris* is very suitable for outdoor mass cultivation because it is resistant to high and low salinity and tolerates sea water and waste water mixtures.

The small cell size (2-6 μm diameter) is however, a disadvantage (Witt et al. 1981). *Nannochloropsis limnetica*, a fresh-water member of Eustigmatophyceae was proposed as fish feed for aquaculture (Krienitz & Wirth, 2006). This alga was capable of achieving 5.5% DW of EPA, 2.2% DW of linoleic acid and 1.05% DW of arachidonic acid.

Nutrient Additives from Algae

PUFAs, especially EPA are recognized as extremely valuable nutritional supplements for man, improving the status of many conditions including heart disease, mental problems, immunological conditions etc. (Bockow et al. 1998, Calder 1999). The following Bacillariophyceae have been identified as potential producers of PUFAs; the marine diatom *Phaeodactylum tricornutum* (Ibáñez 1998; Patil et al. 2007; Meiser et al. 2004), *Monodus subterraneus* (Cohen 1994; Liu & Lin 2005; Lu et al. 2001), and *Nannochloropsis,* both eustigmatophycean algae (Ning et al. 2000), the red alga *Porphyridium cruentum* (Cohen & Cohen 1991), *Pavlova lutheri* from Prymnesiophyceae (Carvalho et al. 2009; Meireles 2003; Patil et al. 2007), and the cyanobacterium *Arthrospira* ssp. (*Spirulina*) (Dilov 1985; Fournadzhieva et al. 2002a). More detailed information is available in the review by Cohen (1999, Chen & Jiang 2001).

Heterotrophic Cultivation

Heterotrophically cultivated *Nitzschia alba* in a stirred-pot fermentor (Kyle & Glaudule 1993), is capable of achieving a DW of 40-50 g. l^{-1} (the autotrophic culture of Bacillariophyceae only achieved 0.2-1 g. l^{-1}) with 40-50% DW of oils. Advantages of cultivation in fermentors include independence of weather and the ability to maintain temperature and light (in mixotrophic mode) at optimal levels.

Trachydiscus Minutus, Strain Lukavský et Pribyl 2005/1

The alga *Trachydiscus minutus* (Bourrelly) Ettl 1964, Xanthophyceae, Heterokontophyta, was found in a small pond in a botanical garden in France, and was described under the name *Pseudostaurastrum minutum* Bourrelly 1951. The alga is wide-spread in the temperate zone, usually present at low concentrations amid the mass of other phytoplankton (Amarasinghe et al., 1997; Zalocar et al., 1998); it is also very rare in Lake Svet, near Třeboň, Czech Rep. (author). The alga that we have studied was isolated from an open shallow cooling pool of the Temelín Nuclear Power Station, Czech Republic. The locality is characterised by a relatively constant temperature of 18-22 ^{0}C year-round, the application of biocides that reduce bacteria and algae, and the addition of $ZnCl_2$ as an anti-corrosive substance; no increased radioactivity was detected. In this locality *T. minutus* dominated through years 2005-6, but it is very rare at present. *Trachydiscus* has often been mistaken for *Chlorella,* or other green spheres (GS), because of a morphological likeness (Petkov & Garcia 2007). The fatty acid composition of the alga, being very different from that of *Chlorella*, is a reliable taxonomic marker, together with lack of chlorophyll_*b* and chl._*c*. The taxonomic position of the genus *Trachydiscus* (*Pseudostaurastrum*) was questioned. Its original place in Xanthophyceae (Heterokonta) was changed into Eustigmatophyceae by Hegewald et al. (2007).

Schnepf et al. (1996) did not find chlorophyll_*c*, in *Pseudostaurastrum limneticum* which confirms the taxonomic membership of this genus in Eustigmatophyceae. The absence of chlorophyll_*c* in *T. minutus* was demonstrated by Přibyl et al. (2012).

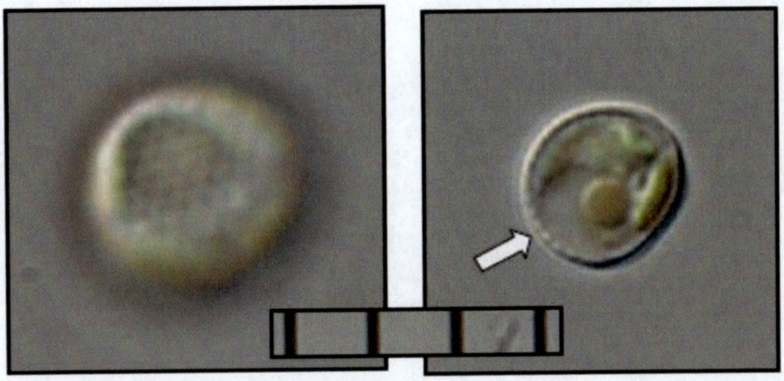

Figure 1. *Trachydiscus minutus* under a light microscope. Characteristic features are a regularly dotted cell wall (left) and a big, centrally placed, yolk-yellow oil drop (right), arrow – openings in cell wall. Scale: 1 unit = 10 μm. After Lukavský et al. (2010).

ULTRASTRUCTURE

Adult *T. minutus* cells are rounded, discoid - pentagonal - irregular, with rounded ends, sometimes with lobes (Fig. 3,4). The diameter is 7-9 μm (>50 – giant cells), with 1 nucleus, 3-4 disc-shape chromatophors, with no pyrenoid. Usually one big, yolk-yellow oil droplet, about 2 μm in diameter, is located in the centre of the cell. The cell wall possesses regular lines of distinct hollows, crossed at 60° (Fig. 1). This structural feature is visible under the light microscope, but only in some cells of the population; there is no clear trend in the appearence of the hollows, e.g. age of cells. Chromatophores are disc-shaped, diameter about 3 μm, parietal, without a peripheral DNA ring or girdle lamella and with no connection to the nucleus. The nucleus sometimes shows condensed chromatin and lamellate vesicles (Fig. 2).

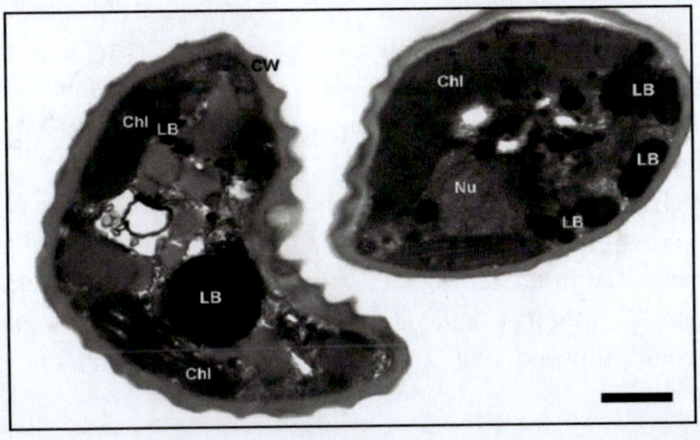

Figure 2. Ultrastructure of *Trachydiscus minutus* cells in TEM. Chl=chloroplasts with thylakoid lamellae, CW=cell wall, LB=lipid bodies, Nu=nucleus. Scale, 1 division = 1 μm. (Přibyl pers. com.).

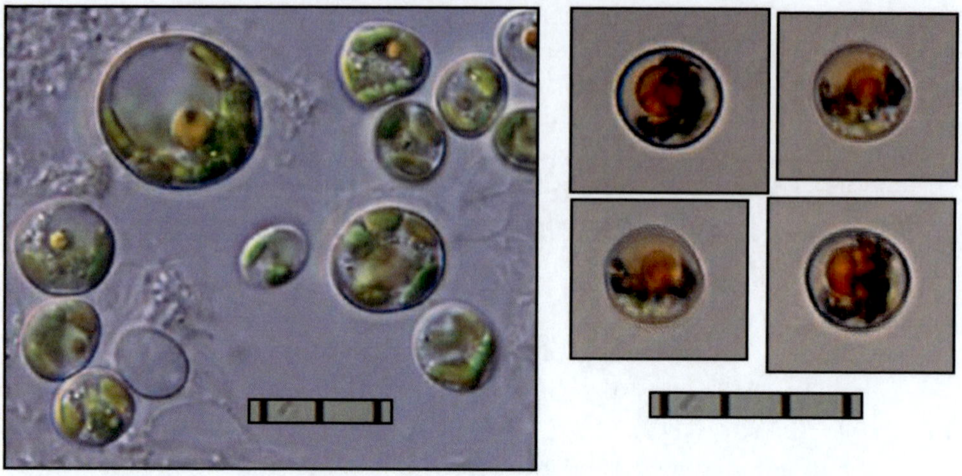

Figure 3. *Trachydiscus minutus* under light microscope. Note the giant cell, with central oil drop (left). Very old cells with deep-red oil droplets with inner inclusions (right). Scale, 1 division = 10 μm. (Orig. author).

Giant cells (Fig. 3 left, Fig.4) are not rare in populations of common cells. They stop their division but continue growth (Fig. 13 arrow). They are not correlated with a high irradiation density, and are also present at low light intensities (Fig. 4).

Photosynthetic Pigments

The principal pigments are violaxanthin (16.2%), then vaucheriaxanthin ester (7.4%), ß–carotene (2.8%), zeaxanthin (0.9%), neoxanthin (1.3%). Only chlorophyll_*a* was detected (71.4% of all pigments). This confirms the taxonomic position of *T. minutus* in Eustigmatophyceae (Přibyl et al. 2012).

Phylogeny

The position of *T. minutus* was in doubt for some time. Now it is based on 18S rRNA gene sequences (Přibyl et al. 2012). The nearest sequence is *Goniochloris sculpta*, with 96% identity and, together with *Pseudostaurastrum enorme* and *P. limneticum,* form a clade with maximal statistical strength.

This clade is a sister to "traditional" eustigmatophyceans; together both clades are in the monophyletic class Eustigmatophyceae. Also the lack of chlorophyll_*c* and the composition of accessory pigments is in concordance. Further research is necessary to resolve the relationship between *T. minutus* and *Goniochloris sculpta*, particularly with regard to Bourrelly (1968) who proposed that *Trachydiscus* resides within the genus *Goniochloris* (Přibyl et al. 2012).

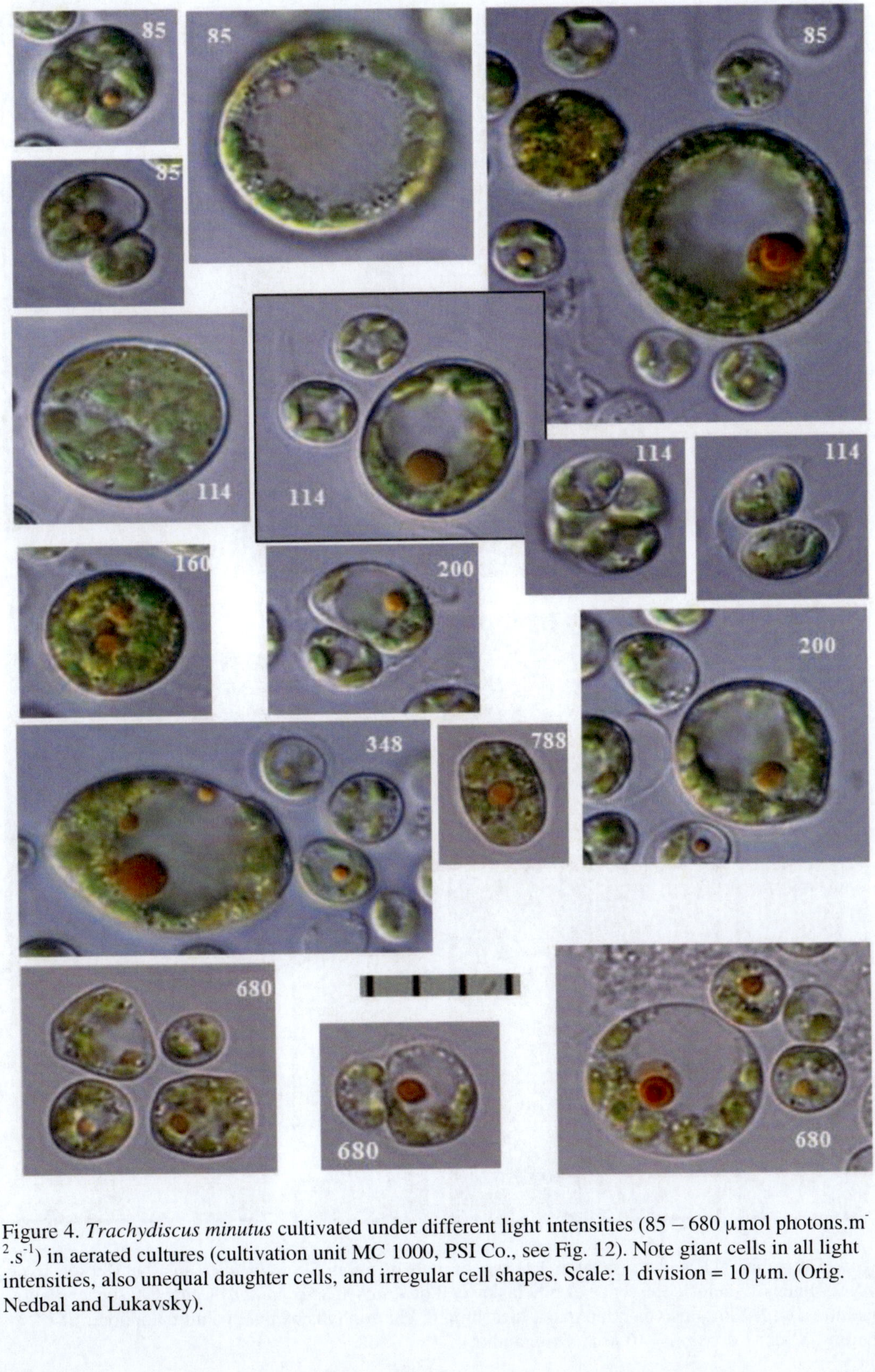

Figure 4. *Trachydiscus minutus* cultivated under different light intensities (85 – 680 μmol photons.m^{-2}.s^{-1}) in aerated cultures (cultivation unit MC 1000, PSI Co., see Fig. 12). Note giant cells in all light intensities, also unequal daughter cells, and irregular cell shapes. Scale: 1 division = 10 μm. (Orig. Nedbal and Lukavsky).

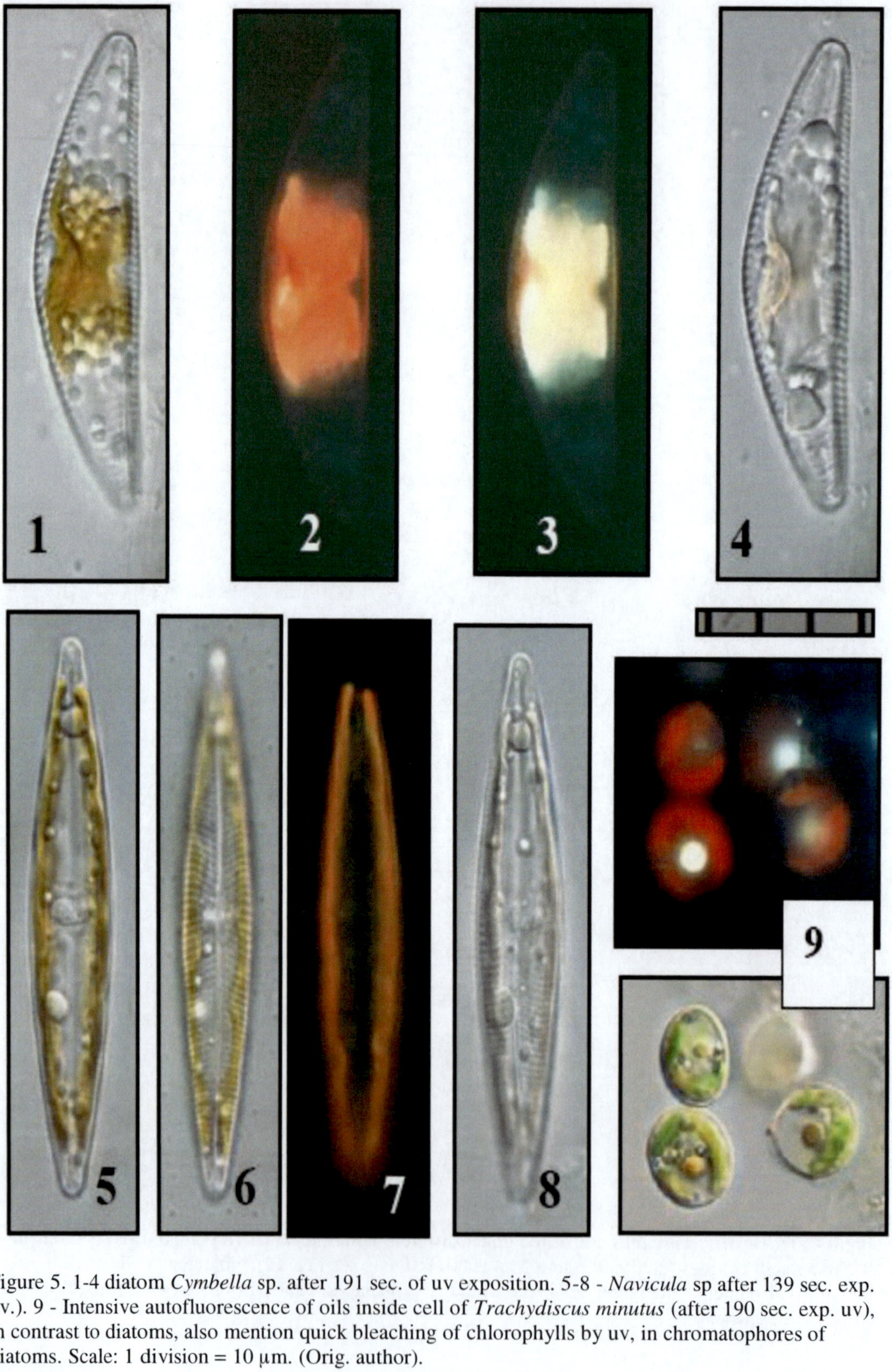

Figure 5. 1-4 diatom *Cymbella* sp. after 191 sec. of uv exposition. 5-8 - *Navicula* sp after 139 sec. exp. uv.). 9 - Intensive autofluorescence of oils inside cell of *Trachydiscus minutus* (after 190 sec. exp. uv), in contrast to diatoms, also mention quick bleaching of chlorophylls by uv, in chromatophores of diatoms. Scale: 1 division = 10 μm. (Orig. author).

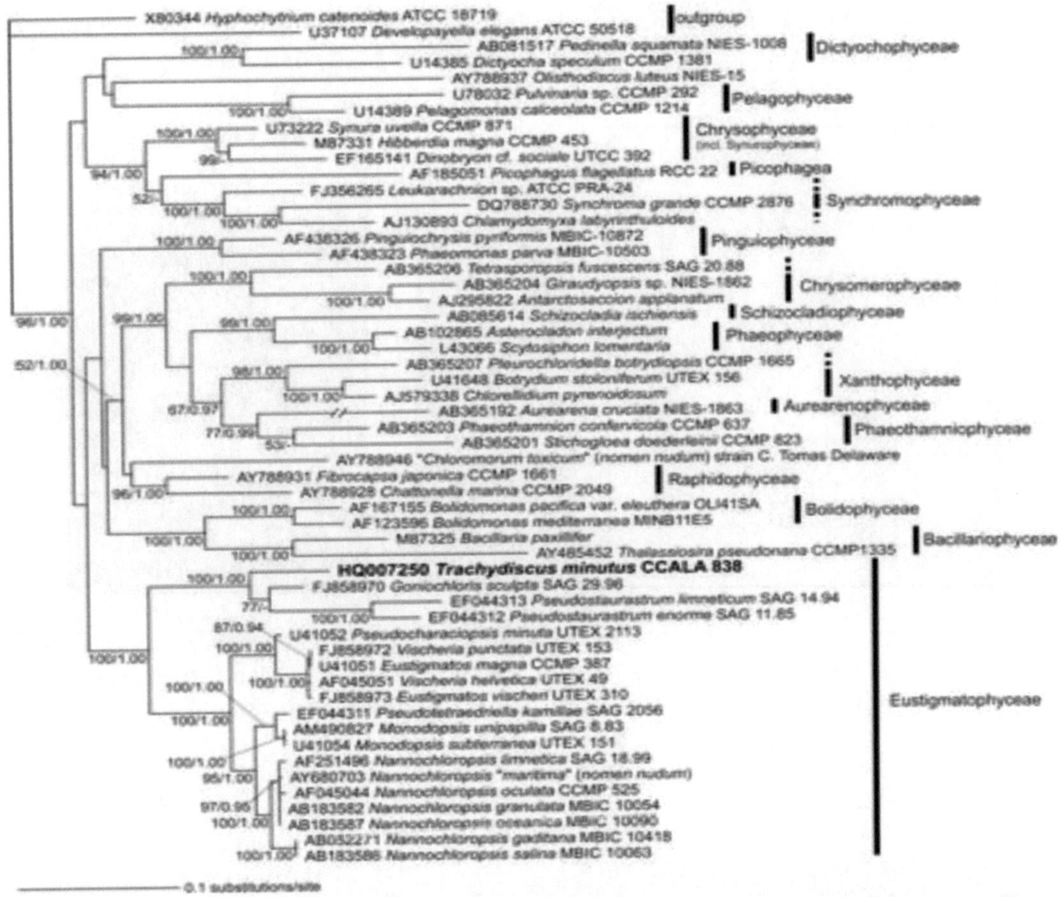

Figure 6. The phylogenetic position of *Trachydiscus minutus* based on the 18S rRNA gene sequence. The ML tree was inferred using RAxML (employing the GTR+Γ+I substitution model). Numbers at branches represent bootstrap support values from the RAxML rapid bootstrapping search/Bayesian posterior probabilities (only values > 50/0.90 are shown). Labels at terminal leaves comprise a DDBJ/EMBL/GenBank accession number of the sequence, source organism, and a strain number (if known). The terminal branch leading to *Aurearena cruciata* NIES-1863 (AB365192) was reduced by half to fit it into the picture. Vertical lines on the right delimit individual heterokontophyte classes as accepted at present. Possible expansion of Synchromophyceae, Chrysomerophyceae and Xanthophyceae to embrace species currently placed elsewhere is indicated by dashed vertical lines. *Olisthodiscus luteus* NIES-15 (AY788937) and "*Chloromorum toxicum*" strain C. Tomas Delaware could not be placed with confidence into any class. *Hyphochytrium catenoides* ATCC 18719 (X80344) and *Developayella elegans* ATCC 50518 (U37107) are treated as an outgroup. After Přibyl et al (2012).

Reproduction

T. minutus usually reproduces via 2-4 autospores (aplanospores). Under specific conditions (dark, temperature about 20 °C) approximately 30% of cells produced zoospores (ZSP) (Fig.7, 8a). These have not yet been observed in nature. ZSPs are elongated, 3.5 x 8.5 - 11 μm, naked, usually with 2 plastids, without eyespot, no pyrenoid, 2 flagella, but only one is easily visible under a light microscope; it is the length of the cell body, with mastigonemes.

The short flagellum is thick and hardly visible (Fig. 7c); movement is slow. Zoosporogenesis is inhibited by N starvation, in contrast with other algae eg. Chlamydomonas, Scenedesmus obliquus, Oedogonium etc. Sexual reproduction has not yet been observed.

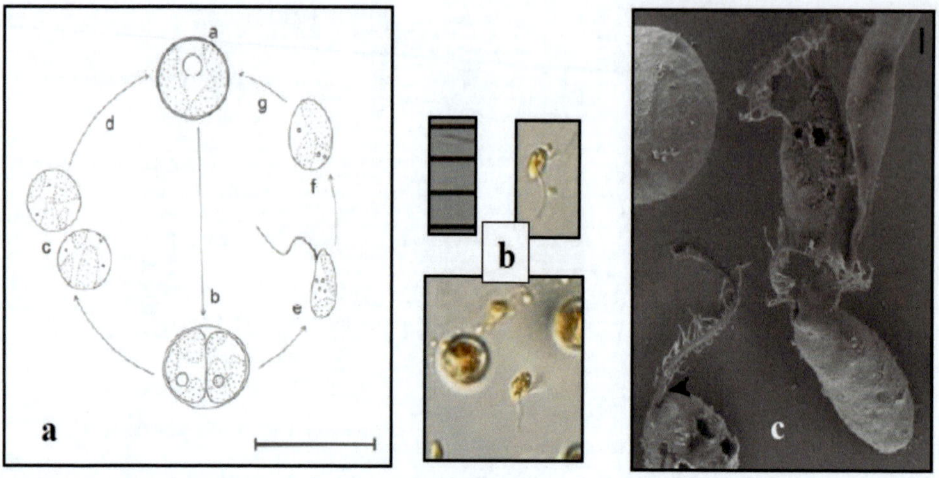

Figure 7. a - Life cycle of *Trachydiscus minutus* (left). b - zoospores under a light microscope, after staining with Lugol solution, and c - under SEM. Arrow - second, small flagellum. Mastigonematas emerging the main flagellum. a, c - after Přibyl et al. (2012), b - Orig. author. Scale: 1 division = 10 μm.

GROWTH CHARACTERISTICS

Data from laboratory cultivation experiments are not directly applicable in outdoor mass cultivation, where the principal regulators (light, temperature) are oscillating. Nevertheless, this is the first and obligatory step in domestication of every alga.

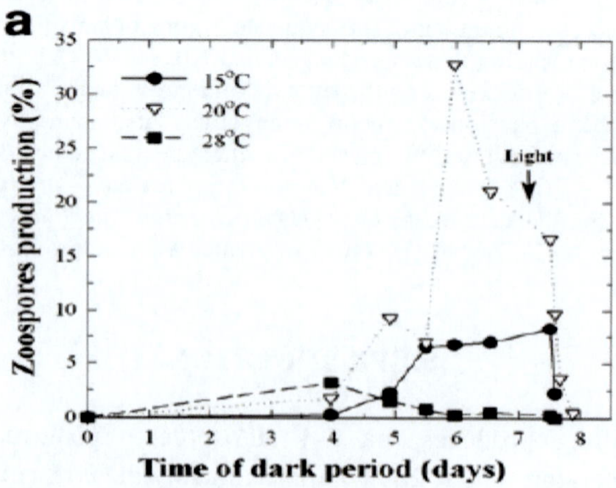

Figure 8. (Continued).

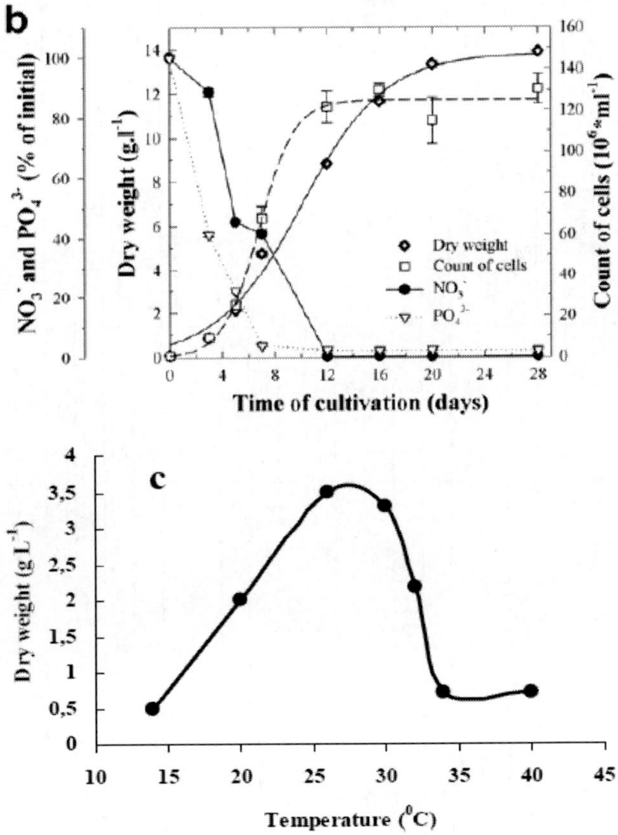

Figure 8. Growth characteristics of *Trachydiscus minutus* under laboratory conditions. a - Time course of zoospore induction in *T.minutus,* using dark cultivation under three different temperatures. Arrow shows time of transfer of cultures onto the light. b - Growth of asynchronous cultures of *T. minutus* and elimination of main nutrient ions (N, P) from the medium during growth. After Pribyl et al. (2012). c - temperature curve after 96 h. of cultivation in laboratory unit, under a light intensity 161 µmol photons.m^{-2}.s^{-1} (PAR). The optimal temperature is about 27 °C. (After Iliev et al. 2010).

In a laboratory unit (Fig. 8b), *T. minutus* showed a standard sigmoidal growth curve and stationary phase was reached at cca 12-14 g DW/L. Depletion of N and P paralleled growth. The temperature optimum was about 27 °C; at higher temperatures, growth fell sharply, but cells were still alive below 40 °C (Fig. 8c).

Biochemical Composition

In *T. minutus,* lipids reached relatively high values; of 39% DW (Table 1), and proteins 48% DW. These proportions characterize *T. minutus* as an alga possessing valuable biomass with high protein and lipid contents. The lipids have been extracted using a mixture of ether/ethanol (Bligh & Dyer 1959), and their odour is similar to fish oil. By TLC analysis, the total lipids comprised triacylglycerols (TAG), monogalactosyldiacylglycerols (MGDG), sulphoquinovosyldiacylglycerols (SQDG), and phosphatidylglycerols (PG). The TAG content of *T. minutus* lipids is about 20 %, which is high.

Table 1. Contents of lipids and their components in algae and cyanobacteria tested until now. TAG - triacylglycerols, EPA = eicosapentaenoic acid. MYR = myristic acid. DW = dry weight. * Cultivated in nutrient deficient medium. After Lukavský et al. 2010, modified

Organism	Lipids	TAG	MYR	EPA	Author
	% in DW	% in DW	% in lipids	% in lipids	
Chlorella protothecoides	53				Cheng et al 2009
Navicula sp.	17-18				Řetovský 1946
Spirulina sp.	9.4				Arvanitis et al. 2003
Spirulina sp.	16.3				Ramadan et al. 2008
Botryococcus braunii	19.8				Shen et al. 2008
Bacillariophyceae	10				Harder et Witch 1942
Scenedesmus obliquus	12.7				Mandal et Mall
12 strains of Cyanobacteria	8-13				Vargas et al. 1998
Nitzschia ovalis			3.1	26.7	Pratoomyot et al. 2005
Thalassiosira sp.			6.4	16.6	
Tetraselmis sp.			1.2	4.7	
Synechocystis sp.			28.2	-	
Synechococcus sp.			26.7	-	
Anacystis sp. (*Synechococcus nidulans*)			4.9	-	
Phaeodactylum tricornutum				5	Meiser et al. (2004)
Nannochloris sp.				23	Petkov et al. (1994)
Nannochloropsis sp.				25	Ning Zou et al. 2000
Nannochloropsis limnetica			6.3	28-56	Krienitz et Wirth (2006)
Pavlova lutheri, UV mutant				32.8	Meireles 2003
Monodus subterraneus				31-34	Liu et Lin 2005
Monodus subterraneus	11			31.8	Lu et al. 2001
Monodus subterraneus				34.2	Vazhappilly et Chen, 1998
Chlorella minutissima				31.3	
Phaeodactylum tricornutum				21.4	
Monodus subterraneus				36.7	Cohen 1994
Trachydiscus minutus	26-31	20	26	27-42	Iliev et al. 2010
*Trachydiscus minutus**	28-39		27-53	10-36	Řezanka et al. 2010

Table 2. ***Trachydiscus minutus*** **cultivated in the laboratory and in a pilot-plant unit in a green house**

Fatty acids.	Total lipids (% of wet weight)				Triacylglycerols
	83 (μmol photons.m^{-2}.s^{-1})	161 (μmol photons.m^{-2}.s^{-1})	[a] 161 (μmol photons.m^{-2}.s^{-1})	[b] 294 (μmol photons.m^{-2}.s^{-1})	161 (μmol photons.m^{-2}.s^{-1})
12:0	tray	0.8 ± 0.4	1.2 ± 0.5	0.6 ± 0.5	1.2 ± 0.1
14:0	28 ± 6	28 ± 3	26 ± 5	4 ± 2	20.3 ± 0.1
16:0	10 ± 2	8 ± 1	9 ± 4	33 ± 13	13.8 ± 0.9
16:1	5 ± 1	10 ± 3	10 ± 3	6 ± 1	14.8 ± 0.9
16:2	tray	0.8 ± 0.3	1.5 ± 1	4 ± 2	0.8 ± 0.1
18:0	0.4 ± 0.1	0.4 ± 0.2	0.6 ± 0.5	3 ± 1	3.7 ± 0.4
18:1	3 ± 1	3 ± 0.5	3 ± 2	9 ± 3	9.9 ± 0.8
18:2	5 ± 1	7 ± 1	7 ± 1	23 ± 7	11.9 ± 0.5
γ-18:3	0.2 ± 0.1	0.8 ± 0.4	0.6 ± 0.4	0.3 ± 0.3	trace
α-18:3	2 ± 1	7 ± 2	5 ± 3	6 ± 2	5.5 ± 0.3
20:4	4 ± 1	6 ± 1	6 ± 2	4 ± 2	10.7 ± 0.9
20:5	42 ± 6	27 ± 4	29 ± 7	3 ± 1	7.4 ± 0.8

[a] arithemtic means and standard deviations from 8 replicates in the laboratory, under 20, 26, 30, 34 $^{\circ}$C, 161 μmol photons.m^{-2}.s^{-1} , [b] cultivation in an orbicular pilot-plant in a greenhouse, Nigrita, Greece, with an area of 2 m^{2} and a volume of about 330 litre, with continuous stirring using a paddle wheel. CO_2 was supplied in excess by bubbling from a perforated tube. Irradiation density was 1100-1200 μmol photons.m^{-2}.s^{-1} in PAR, the temperature of suspension was 26 ^{0}C. Myristic acid = 14:0, γ –linoleic acid = γ-18:3, EPA = 20:5. Content of individual oil acids are calculated from total lipids or triacylglycerols. According Iliev et al. (2010).

Table 3. Composition (% of DW) of lipids in *Trachydiscus minutus* cultivated in nutrient solution after Zachleder et Šetlík (1982) with different concentrations of S, N, and their combinations. DW increase is the change after a starting concentration of 6.2 g.l^{-1}. Irradiation density was 450μmol photons ·m^{-2}·s^{-1}, provided by fluorescent tubes (daylight), bubbled with air enriched with 2% of CO_2, temperature 12-15 °C. After Rezanka et al. 2010

Component		Dist. water	Medium - S	Medium –S, -N	Medium +S, +N
Total lipids		28	33	39	30
Fatty acids	14:1	26.6	41.8	47.1	53.5
	16:1ω-7	11.6	13.0	10.5	8.6
	16:0	11.7	17.5	17.2	15.0
	18:2 ω-6	6.7	6.7	5.4	2.8
	18:2 ω-3	5.9	10.7	7.7	3.8
	20:4ω-6	1.7	0.6	0.6	0.9
	20:5ω-3	35.8	9.7	11.5	15.4
DW increase		0.6	1.0	1.0	5.7

The percentage of fatty acids depends on culture conditions, especially irradiation intensity and nutrient concentrations, and their values under optimum growth conditions are shown in Table 2. A high percentage of MYR (myristic acid) (28 %) and EPA (27-42% W/W-wet weight), was found (Iliev et al. 2010). A relatively high level of sterols was found, the main sterol being β-sitosterol (61.5 %, Iliev et al. 2010). The volatiles of *Trachydiscus,* comprising 0.2 % of the dry biomass, were gravimetrically evaluated and identified by GC/MS. From the total content of oils, myristic acid is 26 % W/W and EPA was up to 42 % W/W. They are both valuable products used in cosmetics (myristic acid) or as food supplements for people, animals and for aquaculture. Also γ -linoleic acid (GLA) has numerous applications. As is clear from Table 2, the EPA content is very much dependent upon cultivation conditions. *T. minutus* is characterised by a high content of both acids, where myristic acid is saturated and EPA unsaturated. Usually in algae, only one prevails; myristic acid in cyanobacteria and EPA in Bacillariophyceae (Table 1). Under different nutritional conditions, the biomass composition of *T. minutus* differs greatly. (Table 3, 4). The most interesting product, EPA, comprises 9.7-35.8% of DW. This enables us to control the composition of the final product according to requirements. *T. minutus* rapidly grows in standard mineral nutrient solutions common in algal biotechnology. Biomass can reach 15 g of dry weight/litre (Fig. 8), and harvesting by sedimentation followed by centrifugation is very easy. In a thin, dense layer of suspension (Fig. 15b) DW as much as 30 g.l^{-1} was reached (Cepák, pers.com.) which is the same as in heterotrophic cultivation. The temperature optimum is about 27 °C, which is reached spontaneously in suspension, without any necessity for expensive temperature control (Fig. 8c). For this reason the alga is suitable for cultivation under temperate as well as subtropical climatic conditions. The biomass contains about 48 % DW of proteins and is a potential feed resource for man or animals. Different combinations of temperature and light resulted in differences in the composition of biomass, pigments, sugars, lipids, proteins, some antioxidants, and proteolytic enzymes. *T. minutus* tolerates temperatures from 20- 32 °C, where growth differences are only moderate. Under unfavourable temperatures of 15 or 40 °C there was a decrease in total lipids to 35% DW, a decrease in carbohydrates to 14% DW, and total proteins were 30-38% DW. Defense systems reacted in two ways: At 40 °C superoxide dismutase, catalase and proteases were elevated and glutathione reductase and peroxidase were decreased. At 15 °C all enzymes except glutathione reductase were inhibited. Light enhanced/suppressed the temperature responses, and chlorophyll and RUBISCO decreased at both temperature extremes. The alga has broad adaptability and resistence to extremes of temperature and light (Gigova et. al. 2012).

Table 4. Variation in the end-point growth (mg DW), protein, carbohydrate, lipid and chlorophyll content (% DW) of *Trachydiscus minutus* as a function of growth temperature and light intensity. Light 132 μmol photons $\cdot m^{-2} \cdot s^{-1}$ unilateral illumination; light 2x 132 μmol photons $\cdot m^{-2} \cdot s^{-1}$ bilateral illumination. All values are expressed as means ± SD; n = 6. After Gigova et al. (2012)

Intensity (μmol photons $m^{-2} \cdot s^{-1}$)	Temperature (°C)				
Light 132	15	20	25	32	40
Dry weight ($mg \cdot ml^{-1}$)	0.75±0.07	2.23±0.13	3.34±0.17	2.93±0.15	0.96±0.12
Total proteins	38.13±2.18	39.76±3.44	38.20±2.01	38.04±1.41	34.96±1.97
Total carbohydrates	13.95±1.25	22.78±1.51	23.74±1.16	23.00±1.44	17.45±2.28
Total lipids	28.19±2.13	22.76±1.95	23.22±2.40	26.66±1.13	30.08±1.54
Chlorophyll_*a*	1.04±0.01	1.08±0.03	1.44±0.08	1.77±0.09	0.65±0.11
Light 2x132					
Dry weight (mg. ml^{-1})	0.79±0.07	2.61±0.15	4.63±0.28	4.03±0.17	0.94±0.15
Total proteins	30.51±1.03	36.57±3.08	33.45±2.99	36.49±3.36	35.38±3.73
Total carbohydrates	13.64±0.85	20.98±0.83	29.01±0.70	25.13±1.04	17.37±2.43
Total lipids	35.19±0.97	23.31±1.48	21.77±1.82	23.27±1.50	27.08±3.92
Chlorophyll_*a*	1.02±0.03	0.92±0.08	1.12±0.01	1.16±0.09	0.73±0.05

INCREMENTS OF POLYUNSATURATED TRIACYLGLYCEROLS

The content of TAGs is dependent on the composition of the nutrient medium. Seventy four TAGs having at least one PUFA were determined in *T. minutus* (Řezanka et al. 2010, 2011). Palmitic, arachidonic and EPAs are the main components. N and P starved cultures produced TAGs without PUFAs at position sn-2 (Tab. 5). The composition of biomass can be easily controlled via manipulation of nutrients, which is only possible in outdoor cultivation.

Table 5. Ratios of TAG positional isomers consisted of EPA, ARA and palmitic acid in the alga *Trachydiscus minutus*, under different nutrient supply. PEP - 1,3-dipalmito-2-eicosapentaenoin, PPE - 1,2-dipalmito-3-eicosapentaenoin, PEE - 1-palmito-2,3-dieicosapentaenoin, EPE - 1,3-dieicosapentaenoin-2-palmitin, PAP - 1,3-dipalmito-2-arachidonin, PPA - 1,2-dipalmito-3-arachidonin, PAA - 1-palmito-2,3-arachidonin APA - 1,3-diarachidonin-2-palmitin. After Rezanka et al. (2011)

TAG ratio	control	-N	-P
PEP/PPE	1:3	3:1	5:1
PAP/PPA	1:3	3:1	4:1
EPE/PEE	1:10	2:1	3:2
APA/PAA	1:10	2:1	3:2

Visualization of the Oil Drops

The central oil spheres of *T. minutus* are easily visible *in vivo* under the light microscope (Figs 1,3,4). Under epifluorescence microscopy, the cells autofluoresce (Řezanka et al. 2010). The oil is excited under uv (wavelength 360-370 nm) and observed or photographed, using a barrier filter (>515 nm). Oil shows a white colour, chlorophyll appears as red, and cell walls are blue. The oil requires some time to change from yolk-yellow to white or blue-white; about 2 minutes (Fig. 9). This autofluorescence of oils under uv is specific for Xanthophyceae and Eustigmatophyceae. It was not observed in e.g. Bacillariophyceae (Fig. 5), nor in Chlorophyta and Cyanobacteria (Přibyl pers.com.). The oil is synthesized as numerous small drops inside the chromatophore, and passes into the central space to coalesce into one big drop (Přibyl et al. 2012). Under uv microscopy, live cells and dead cells can be distinguished, e.g. in immobilized pellets (Figs. 9, 11). This is potentially a way for quick evaluation of intracellular oils instead of extraction with solvents and gravimetric analysis (Přibyl pers.com). The disadvantage is that chlorophyll as well as cell structures are soon damaged by intensive uv. Interestingly *T. minutus* (which only possesses chlorophyll_*a*) is more resistant to uv than the procaryotic cyanobacteria and eucaryotic algae.

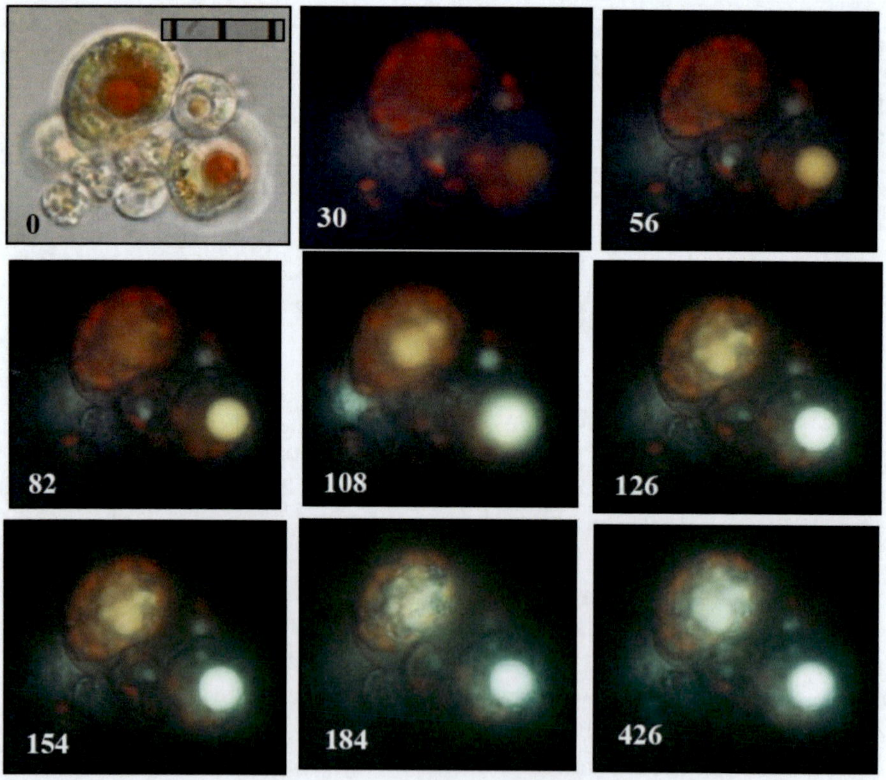

Figure 9. Autofluorescence of oil drops and chlorophyll of *Trachydiscus minutus* under uv, over a period of 30-426 sec. Cells were immobilized in 2% agar in Petri dishes. 0- visual light. Continuous bleaching of chlorophyll is visible, too. A cover glass is placed on the surface of the agar plate to use an immersion objective HI 100x. Scale: 1 division = 10 μm. Orig. author.

Raman Microspectroscopy

Oils produced by *T. minutus* are characterised by a high degree of unsaturation, which is usually expressed as the iodine index. This crucial parameter can be evaluated directly inside cells, in the oil drops, by Raman spectroscopy (Fig. 10), after calibration with known standards. The iodine index differs between algal species: *Chlamydomonas* sp. - 93, *Botryococcus sudeticus* (*Botryosphaerella sudetica*) - 88 and *T. minutus* - 216. The differences in intensities of Raman spectral peaks can be exploited in the selection of novel algal strains as well as in the production of new mutants. *T. minutus* has potential as a food supplement, but not for bio-diesel, where an iodine value of about 120 is required (Samek et al. 2010).

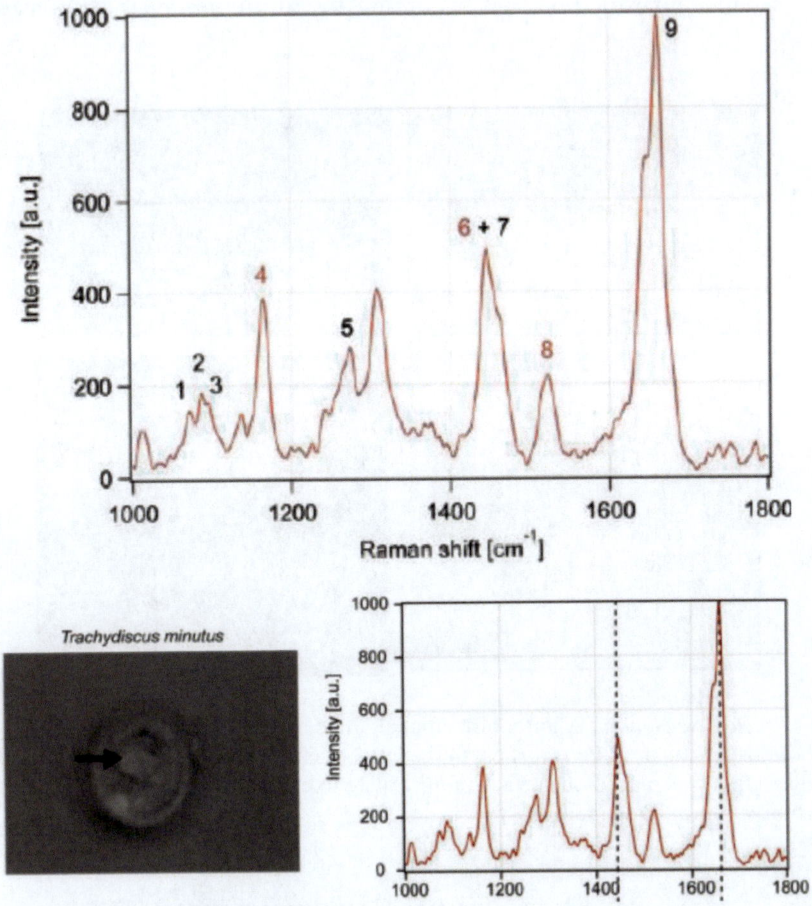

Figure 10. a - Raman scattering spectrum of *Trachydiscus minutus*. Raman bands 7 and 9 are used to calculate the degree of lipid unsaturation. Spectrum acquisition parameters: integration time 20 sec., laser power at the specimen 13 mW. 4,6,8 numbers indicate β - carotene vibrations. b - Typical Raman scattering spectra of intracellular lipid bodies contained in *Trachydiscus minutus*. Raman bands used for the calculations of iodine value are highlighted with dashed vertical lines. Corresponding pictures to the left of the spectra show the lipid bodies from which the spectra were recorded (arrow). After Samek et al. (2010).

Pilát et al. (2011) used Raman microspectroscopic analysis for assessment of β-carotene concentration in *T. minutus* algal lipid bodies of *in vivo*. This is useful in biotechnology for lipid and pigment production, as it allows rapid selection for overproducing strains.

IMMOBILIZATION

Immobilized (IM) algae characteristically cease growth and cell division after ca 3 divisions, but metabolic activity proceeds for a long time (Lukavský et al.1986). Algae immobilized in agar in test tubes survived for 3 years (Lukavský 1988). *T. minutus* can be readily immobilized in agar or alginate and cells are viable for months. Individual cells divided 2-3 times and produced characteristic mulberry-like colonies (Fig.11). The oil content in IM cells is under investigation, but according to autofluorescence responses, should be high.

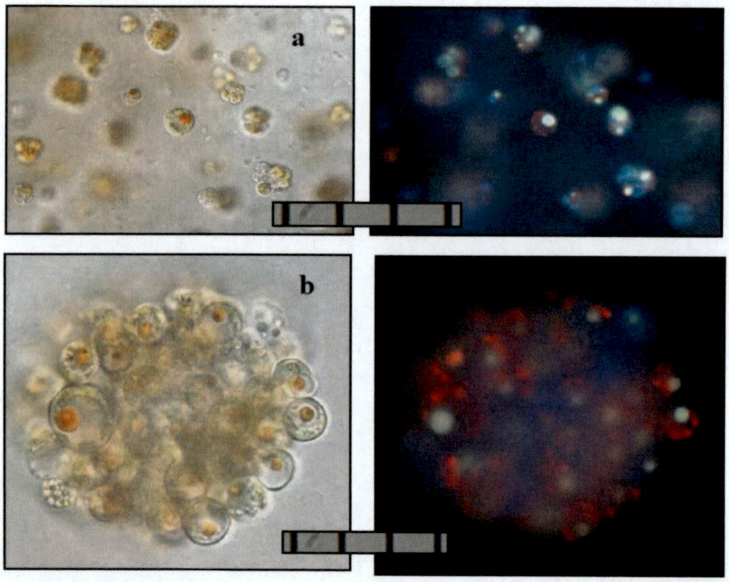

Figure 11. a - *Trachydiscus minutus* immobilized in alginate, squash preparation. b - *Trachydiscus minutus* immobilized in agar, in Petri dishes. In the left – visible light, in the right – autofluorescence under uv. Cover glass was placed on surface of agar plate to enable using immersion objective HI 100x. Scale: 1 division = 10 μm. Orig. author.

LABORATORY CULTIVATION

T. minutus can be easily cultivated in laboratory units, in suspensions (Fig. 12) as well as on agar plates (Fig. 13), and can be immobilized in agar or alginate (Fig. 9, 11). Selection of the proper technique and unit depends on the aim of the experiment. Selection for growth mutants is possible by evaluation of colonies and their isolation in multiwell immunological plates (Fig. 14). Almost 100 mini-cultures (0.25 ml) can be exposed and evaluated quickly in a plate reader, at 750 nm, and the data can be converted into DW. Testing of growth parameters is optimal in test tubes in a water bath; greater volumes e.g. for inoculation of outdoor cultivation, or analyses demanding huge biomasses, can be produced in 15 litre vessels (Fig. 12-right).

Cultivation in Petri dishes, as suspensions or colonies on agar plates, in equipment with crossed gradients of temperature and light (Lukavský 1982, Kvíderová and Lukavský 2001) can allow ready evaluation of growth characteristics, and biomass composition.

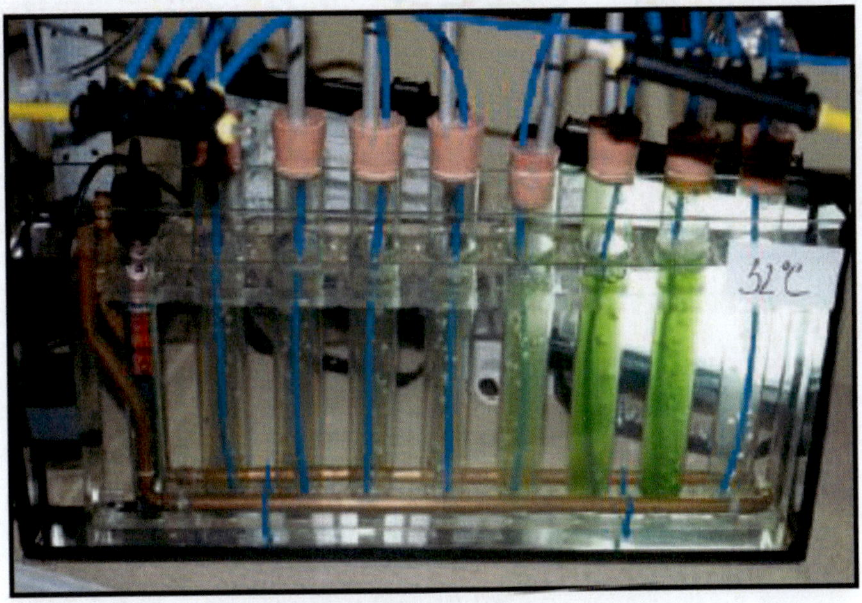

Figure 12. *Trachydiscus minutus* in laboratory cultivation units, in left test tubes in a water bath, irradiated from one side, after Nedbal unpublished). On the right is a plane parallel glass cuvette, of volume up to 15 litre, irradiated from both sides with fluorescent tubes (2x 100 µmol photons.m^{-2}.s^{-1}), aerated by perforated tube, with air containing 2% of CO_2. Orig. author.

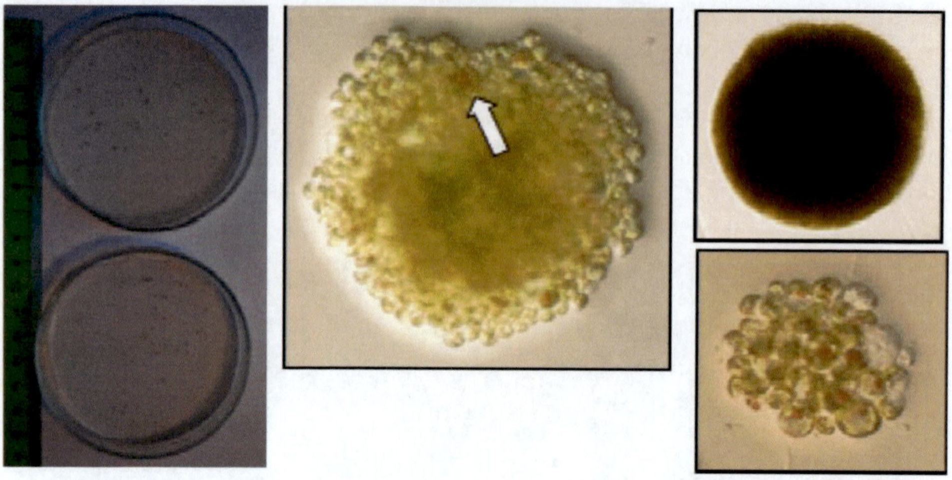

Figure 13. *Trachydiscus minutus* cultivated as single- cell colonies on agar plate in a Petri dish, and details of colonies. Arrow – giant cell. Orig. author.

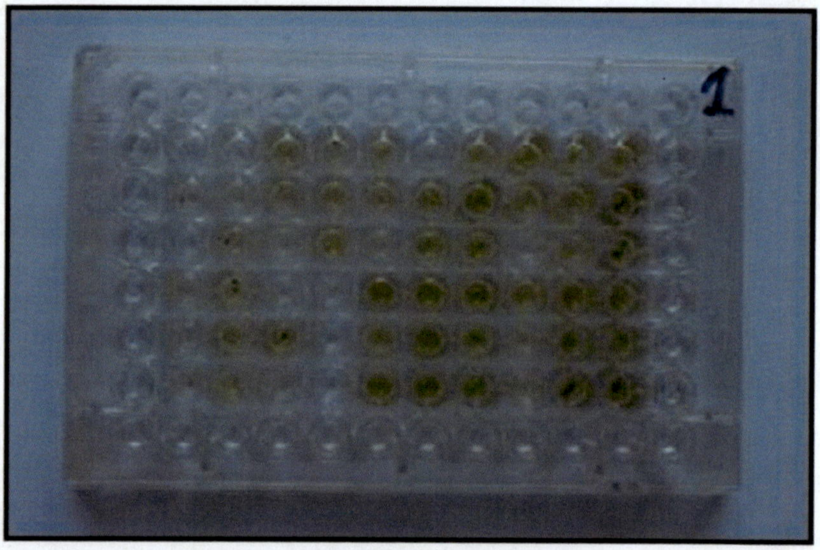

Figure 14. Cultivation of *Trachydiscus minutus* in an immunological plate, FB type, wells of 0.2 ml. The plates can be evaluated directly by plate reader, e.g. OD 750 can be used for plotting of growth curves. Marginal wells are filled with distilled water to diminish evaporation. Orig. author.

T. minutus grows well on agar plates in Petri dishes. This cultivation method is suitable for isolation and purification of algae because it results in bacteria-free clonal cultures. Genetic experiments can also be carried out using solidified nutrient medium. Colonies on agar plates can be used not only for selection of qualitative mutants, e.g. according differencies in colour, but also in the selection of algal clones according to growth rate, e.g. *Chlorella* (Lukavský 1975). *T. minutus* does not grow quickly, but it does yield distinct colonies (Fig. 13). The correlation between colony diameter and optical density of their progeny in stationary cultures is significant (Fig. 16).

PILOT PLANT CULTIVATION

T. minutus cultivated in a 330 litre orbicular, flat vat (Fig. 15a), in a greenhouse, in N. Greece, for 30 days, gave an average daily yield of 7 g DW.m^{-2}, under routine conditions, where the natural light intensity was up to 1100-1200 μmol photons ·m^{-2}·s^{-1} in PAR, the maximum suspension temperature was 26 ^{0}C, and CO_2 was supplied in excess via a perforated tube. Harvesting by centrifugation and drying by sun irradiation on plastic foil was successful. Cultivation as a dense, thin layer was also carried out, on an inclined, flat platform, in a greenhouse at Trebon, SW. Bohemia (Fig. 15b). The yields in both units are compared in Table 6; the thin layer was more effective with respect to volume and area than the orbicular unit.

a b

Figure 15. Cultivation units exploited for pilot-plant cultivation of the alga *Trachydiscus minutus*. a - orbicular unit with suspension layer 20-30 cm high, stirred by paddle wheel, CO_2 is supplied by perforated tube, greenhouse of Algae Co, Nigrita, N. Greece. b - flat, inclined unit with a thin (3-5 mm), dense layer (BSC Co., Brno CZ). Pump services mixing as well as CO_2 enrichment (green house at Třeboň, SW. Bohemia). After Lukavský et al. (2010).

Another advantage of mass cultivation is that *T. minutus* is resistant to fungal parasites e.g. Chytridiales, which are common in cultures of *Scenedesmus* and *Chlorella* (*Phlyctidium scenedesmi*, *Aphelidium chlorococcarum*). No parasites have been observed, yet.

Table 6. A comparison of EPA production by an orbicular unit in Nigrita, N. Greece, and a flat, inclined surface unit in Třeboň, SW. Bohemia. Both units were placed in a greenhouse, without temperature control (see Fig. 15). Original author

	Nigrita	**Třeboň**
Lipids in dry weight (%)	26	28-31
EPA in dry weight (%)	3 ± 1	7.2 – 13.02
Yields of dry weight/m^2/day (g)	19	2.8
Yields of dry weight/L/day (g)	0.115	0.3
Yields of EPA/m^2/day (mg)	0.57	0.25 – 0.455
Yields of EPA/L/day (mg)	3.45	20 – 30
Volume of unit	200	140
Area of unit (m^2)	2	15
Height of suspension (mm)	200 - 300	5 - 10

POTENTIAL FOR INCREASED EPA CONCENTRATIONS

T. minutus cultivated in a standard medium (after Zehnder in Staub, 1961) and in the same solution without S and N, after a 16-d cultivation period, reached 13 g dry mass per 1 litre of medium (Tab. 7). The content of oligoenoic ('polyenoic') fatty acids (PUFAs), i.e. eicosapentaenoic (EPA), was in excess of 35 % of total fatty acids; productivity was thus 88 $mg.l^{-1}.d^{-1}$ (Řezanka et al. 2010).

Table 7. Composition (% DW) of total lipids and fatty acids and dry mass (in $g.l^{-1}$) increase in *Trachydiscus minutus* cultivated under different nutritional conditions. Dry mass found at the onset of stationary phase (6.2 $g.l^{-1}$) before the samples from the mother culture were mixed with the same volume of H_2O, medium (–S), (–S,–N) or full cultivation medium (+S,+N). Light; 450 µmol $photons.m^{-2}.s^{-1}$, from fluorescent tubes (day light); cultures bubbled with air containing 2% CO_2. After Rezanka et al. (2010)

Component (% of dry mass)		Distilled water	Medium - S	Medium –S, -N	Medium +S, +N
Total lipids		28	33	39	30
Fatty acid	14:0	26.6	41.8	47.1	53.5
	16:1ω-7	11.6	13	10.5	8.6
	16:0	11.7	17.5	17.2	15.0
	18:2ω-6	6.7	6.7	5.4	2.8
	18:3ω-3	5.9	10.7	7.7	3.8
	20:4ω-6	1.7	0.6	0.6	0.9
	20: 5ω-3	35.8	9.7	11.5	15.4
Dry mass increase		0.6	1.0	1.0	5.7

POSSIBILITIES OF SCREENING AND SELECTION

Classic genetic breeding and improvements are based on the production of mutants by uv irradiation or chemical mutagens, followed by screening for required clones. A prerequisite for such a procedure is a correlation between diameter of colonies and growth rates, in derived suspensions; this has been proven (Fig. 16). The correlation coefficient r = 0.55 is significant at 1% P (Fig. 16).

An advantage of using colonies on an agar plate, is that individual cells can also be observed and evaluated, e.g. giant cells can be recognized inside a colony (Fig. 13 arrow). Coefficients of correlations are almost identical if only straight line or quadratic regressions are calculated and plotted (Fig. 16). The correlation is, however, supported mainly by the tail of small colonies which gave little growth in suspension. The greatest colonies resulted in a wide spectrum of yields. Further testing is necessary.

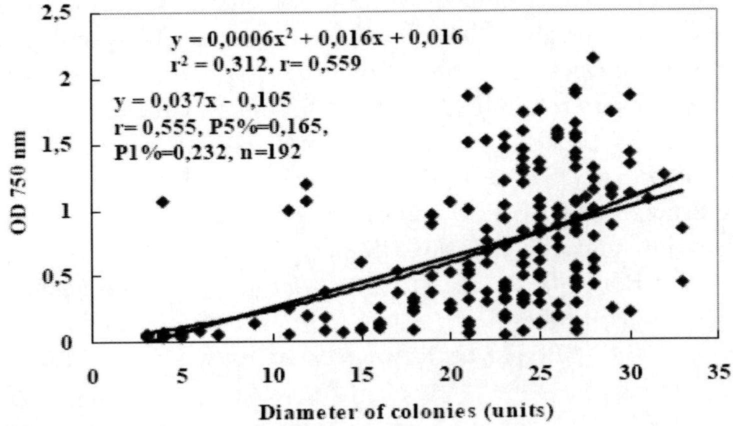

Figure 16. Correlations between diameters of colonies on an agar plate and OD750 nm in stationary phase of 10 ml volume cultures in test tubes. Petri plates as well as tubes were irradiated using fluorescent tubes, 164 μmol photons $\cdot m^{-2} \cdot s^{-1}$ (day light), temperature 25 °C, CO_2 was supplied only by diffusion from air. Original author.

Conclusion

Trachydiscus minutus is shown to be a new organism with great potential for the production of PUFAs, especially of EPA. The strain, Lukavsky et Pribyl 2005/1, was patented in the Czech Republic (Lukavsky et al. 2010). The yields of EPA from T. minutus are comparable, sometimes better, than other algae in use at present.

Acknowledgments

This review would not be posssible without the assistance of many enthusiastic people involved in the study, I thank R.Andreeva, H.Brabcová, Vl.Cepák, M.Eliáš, S. Furnadzhieva, G.Gacheva, L.Gigova, I.Iliev, N.Ivanova, P.Kaštánek, L.Nedbalová, P.Novák, M.Pažoutová, G.Petkov, J.Petrů, P.Přibyl, T.Řezanka, K.Sigler, M.Vosátka, V.Zachleder, M.Zulumidis and Algae a.s. The project No.TE 01020080, Biorefinery Research Centre of Competence, Technology Agency of The Czech Republic, for financial support. I also thank editors of the journals for their kind permission to copy some figures and graphs (Elsevier, Willey, Folia Microbiol., Sensors and Gen.Appl. Plant Physiol.) Last but not least, J.D. Brooker who proof-read the manuscript.

References

Amarasinghe, P.B, Boersma, M. & Vijverberg, J. (1997). The effect of temperature, and food quantity and quality on the growth and development rates in laboratory-cultured copepods and cladocerans from a Sri Lankan reservoir. *Hydrobiologia,* 350, 131-144.

Arvanitis, A., Sotiroudis, T.G, Nerantzis, E., Fournadzhieva, S. & Koultsiakis, E. (2004). Mass culture of the microalga *Spirulina* using geothermal fluids in Greece. Antioxidant activities of *Spirulina* powder extracts. In *Proc. Internat. Conf. "Geothermal Energy Applications in Agriculture"*, 3-4/5/2004, pp. 1-6, Athens, Greece.

Bligh, E.G. & Dyer,W.J. (1959). A rapid method of total lipid extraction and purification. *Can. J. Biochem. Physiol.* 37, 911–917.

Bockow, B.L. & Erlitz, M.D. (1998). Compositions of *Spirulina* algae and omega fatty acids for treatment of inflammation and pain. US Patent No. 5,709,855.

Borowitzka, M. & Borowitzka, L. (1988). *Micro-algal Biotechnology*. Cambridge, Cambridge University Press.

Borowitzka, M.A. (1988). Microalgae as sources of essential fatty acids. *Australian J. Biotechnol.,* 1, 58-62.

Borowitzka, M.A. (1997). Microalgae for aquaculture: opportunities and constraints. *J. Appl. Phycol.,* 9, 393-401.

Bourrelly, P. (1951). Xanthophycées rares ou nouvelles. *Bull. Mus. National d´hist. naturelle,* 2 sér., 23, 666-672.

Brown, M.R. (2002). Nutritional value of microalgae for aquaculture. In Cruz-Suárez L.E. et al. (Eds.), Advances en Nutrición Acuícola VI. Cancún, Quintana Roo, 2002, Mexico.

Calder, P.C. (2006). n-3 Polyunsaturated fatty acids, inflammation, and inflammatory diseases. *Am. J. Clin. Nutr.,* 83,1505–1519.

Carvalho, A.P, Monteiro, C.M. & Malcata, F.X. (2009). Simultaneous effect of irradiance and temperature on biochemical composition of the microalga *Pavlova lutheri. J. Appl. Phycol.,* 21, 543–52.

Chen, F. & Jiang, Y. (2001). Algae and their biotechnological potential. Kluver, Dortrecht, The Netherlands.

Cheng, Y., Lu, Y., Gao, C.F., & Wu, Q.Y. (2009). Alga-based biodiesel production and optimization using sugar cane as the fedstock. *Energy & Fuels,* 23, 4166-4173.

Cohen, Z. (1994). Production potential of eicosapentaenoic acid by *Monodus subterraneus. J. Amer. Oil Chem.Soc.,* 71, 941-945.

Cohen, Z., & Cohen, S. (1991). Preparation of eicosapentaenoic acid from *Porphyridium cruentum. J. Amer. Oil Chem. Soc.,* 68, 16-19.

Courchesne, N., Lan, Ch., Mu, Y. & Wang, B. (2010). Microalgae for biofuel production and CO_2 sequestration. Nova Publishers, 168pp, NY

Dilov, C. (1985). Mass cultivation and use of microalgae. *Bulgarian Acad. Sci. Publ. House,* Sofia, Bulgaria.

Dousková, I., Doucha, J., Lívanský, K., Machat, J., Novák, P., Umyslová, D., Zachleder, V., & Vítová, M. (2008). Simultaneous flue gas bioremediation and reduction of microalgal biomass production costs. *Appl. Microbiol. Biotechnol.,* 1, 179-185.

Eltgroth, M.L., Watwood, R.L. & Wolfe, G.V (2005). Production and cellular localization of neutral long-chain lipids in the haptophyte algae *Isochrysis galbana* and *Emiliania huxleyi. J. Phycol.,* 41, 1000-1009.

Ettl, H. (1964). *Trachydiscus* nov. gen., eine neue Gattung der coccal Xanthophyceen. *Bot. Notiser,* 117, 213- 221.

Fournadzhieva, S., Pilarsky, P., Arvanitis, A., Fytikas, M., & Koultsiakis, E. (2002a). Use of geothermal fluids for cultivation of the microalga *Spirulina* in Nigrita - Serres. In:. Tsiligiridis & Kaupis (Eds.), Proc. 7th Nat. Conf. Ren. Energy Sources, Patras 6-8.11.2002, *Inst Solar Technol* vol. B, pp. 97-104.

Fournadzhieva, S, Bojadgieva, K, Fytikas, M, & Popovski, K. (2002b). A conception of a geothermal application in common microalgal development in Balkan countries. In: K. Popovski & M. Fytikas (Eds.), International course on district heating, agricultural and agro-industrial uses of geothermal energy. Tessaloniky, 1-4 September 2002, pp 152-159. Aristotle Univ. Tessaloniki, Greece.

Gigova, L., Ivanova, N., Gacheva, G., Andreeva, R. & Furnadzhieva, S. (2012). Response of *Trachydiscus minutus* (Xanthophyceae) to different temperature and light. *J. Phycol.* 48 : 85-93.

Gressel, J. (2008). Transgenics are imperative for biofuel crops. *Plant Science,* 174, 246-263.

Harder, R. & Witsch, H. (1942a). Űber Massenkultur von Diatomeen. *Ber. Dtsch. Botan. Ges.,* 60, 146-152.

Harder, R. & Witsch, H. (1942b). Bericht uber Versuche zur Fettsynthese mittels autotropher Mikroorganismen. *Der Forschungsdienst, Sonderheft* 16, 270-275.

Hegewald, E., Padisák, J. & Friedl, T. (2007). *Pseudoteraëdriella kamillae*: taxonomy and ecology of a new member of the algal class Eustigmatophyceae (Stramenopiles). *Hydrobiologia* 586, 107-116.

Huang, G.H., Chen, F., Wei, D., Zhang, X.W, & Chen, G. (2010). Biodiesel production by microalgal biotechnology. *Applied Energy*, 87, 38-46.

Ibáñez, G. M. J., Medina A. R., Grima, E. M., Giménez, A. G., Carstens, M. & Cerdán, E. (1998). Optimization of fatty acid extraction from *Phaeodactylum tricornutum* UTEX 640 biomass. *J. Amer. Oil Chem. Soc.*, 75, 1735-1740.

Iliev, I., Petkov, G., Lukavský, J., Furnadzhieva, S., Andreeva, R. & Bankova, V. (2010). The alga *Trachydiscus minutus* (*Pseudostaurastrum minutum*): Growth and Composition. *Gen. Appl. Plant Physiology.* 36, 222-231.

Ketchum, B.H. & Redfield, A.C. (1938). A method for maintaining a continuous supply of marine diatoms by culture. *Biol. Bull.,* 75, 165 – 169.

Krienitz, L. & Wirth, M. (2006). The high content of polyunsaturated fatty acids in *Nannochloropsis limnetica* (Eustigmatophyceae) and its implication for food web interactions, freshwater aquaculture and biotechnology. *Limnologica,* 36, 204-210.

Kyle, D. & Glaudule, R. (1993). Eicosapentaenoic acids and methods for their production. US Patent No. 5,244,921.

Kvíderová, J. & Lukavský, J. (2001). A new unit for crossed gradients of temperature and light. In J. Elster et al. (Eds.), Algae and Extreme Environments. Proc. Int. Conf. 11-16. Sept., Třeboň, CZ, Nowa Hedwigia, *Beihefte* 123, 539-548.

Leon-Banares, R., Gonzalez-Ballester, D., Galvan, A. & Fernandez, E. (2004). Transgenic microalgae as green cell-factories. *Trends Biotechnol.,* 22, 45–52.

Li, Q., Du, W., & Liu, D.H. (2008). Perspectives of microbial oils for biodisel production. *Appl. Microbiol. Biotechnol.,* 80, 749-756.

Liu, C.P. & Lin, L.P. (2005). Morphology and eicosapentaenoic acid production by *Monodus subterraneus*. *Micron,* 36, 545-550.

Lomborg, B. (2006). Skeptický ekolog [Sceptical Environmentalist]. Dokořán, Liberální Institut Prague.

Lu, C.M., Rao, K., Hall, D. & Vonshak, A. (2001). Production of eicosapentaenoic acid (EPA) in *Monodus subterraneus* grown in a helical tubular photobioreactor as affected by cell density and light intensity. *J. Appl. Phycol.*, 13, 517-522.

Lukavský, J. (1975). Analysis of growth rate of algae by cultivation on solid media. *Arch. Hydrobiol./Suppl.49, Algolog.Studies,* 14, 105-136.

Lukavský, J. (1982). Cultivation of chlorococcal algae in crossed gradients of temperature and light. *Arch. Hydrobiol./Suppl.60,Algolog. Studies,* 29, 517-528.

Lukavský, J. (1988). Long-term preservation of algal strains by immobilization. In: Progress in Algal Taxonomy, Smolenice, June 15-19, 1987. *Arch. Protistenk.*, Jena, 135, 65-68.

Lukavský, J., Komárek, J., Lukavská, A., Ludvík, J. & Pokorný, J.(1986). Metabolic activity and cell structure of immobilized algal cells (*Chlorella, Scenedesmus*). *Arch. Hydrobiol.,* Suppl. 73, *Algolog. Stud.* 43, 261-297.

Lukavský, J., Cepák, V., Řezanka, T., Petkov, G., Furnadzhieva, S., Iliev, I., Andreeva, R., & Bankova,V. (2010). Řasa *Trachydiscus minutus* kmen Lukavský & Přibyl 2005/1, produkující oleje s vysokým obsahem vícenenasycených mastných kyselin [*Trachydiscus minutus* strain Lukavský & Přibyl 2005/1, alga producing oils with high amount of higher unsaturated fatty acids]. PV 2010-436. Patent č. 302118, *Věstník,* 42/2010.

Mandal, S., & Mallick, N. (2009). Microalga *Scenedesmus obliquus* as a potential source for biodiesel production. *Appl. Microbiol. Biotechnol.,* 84, 281-291.

Mata, T.M., Martins, A.A. & Caetano, N.S. (2010). Microalgae for biodiesel production and other applications: A review. *Renewable and Sustainable Energy Reviews* 14, 217–32.

Meireles, L.A. (2003). Increase of the yields of eicosapentaenoic and docosahexaenoic acids by the microalga *Pavlova lutheri* following random mutagenesis. *Biotechnol. Bioeng.,* 81, 50-55.

Meiser, A., Schmid-Staiger, U. & Trösch, W. (2004). Optimization of eicosapentaenoic acid production by *Phaeodactylum tricornutum* in the flat panel airlift (FPA) reactor. *J. Appl. Phycol.* 16, 215–225.

Narendra, M.V., Mehrothra, S., Shukla, A. & Mishra, B.N. (2010). Prospective of biodiesel production utilizing microalgae as cell factories: A comprehensive discussion. *African J. Biotechnol.,* 9, 1402-1411.

Ning Z., Chengwu, Z., Cohen, Z. & Richmond, A. (2000). Production of cell mass and eicosapentaenoic acid (EPA) in ultrahigh cell density cultures of *Nannochloropsis* sp. *Eur. J. Phycol.,* 35, 127-133.

Patil, V., Kallquist, T., Olsen, E., Vogt, G. & Gislerod, H.R. (2007). Fatty acid composition of 12 microalgae for possible use in aquaculture feed. *Aquaculture Internat.,* 15, 1-9.

Petkov, G. & Garcia, G. (2007). Which are fatty acids of the green alga *Chlorella*? Biochem. *Syst. Ecol.,* 35: 281-285.

Petkov, G., Furnadzieva, S.T. & Andreeva, R.D. (1994). Fatty acid and sterol composition on *Nannochloris* sp. *Algol. Stud.,* 72: 133-135.

Petkov, G., Ivanova, A., Iliev, I. & Vaseva, I. (2012). A critical look at the microalgae biodiesel. *Eur. J. Lipid Sci. Technol.* DOI: 10.1002/ejlt.201100234.

Pienkos, P.T. & Darzins, A. (2009). The promise and challenges of microalgal-derived biofuels. *Biofuels Bioproducts & Biorefining-Biofpr.,* 3, 431-440.

Pilát, Z., Bernatová, S., Ježek, J., Šerý, M., Samek, O., Zemánek, P., Nedbal, L. & Trtílek, M. (2011). Raman microspectroscopy of algal lipid bodies: β-carotene quantification. *J Appl Phycol.,* DOI 10.1007/s10811-011-9754-4.

Pratoomyot, J., Srivilas, P. & Noiraksar, T. (2005). Fatty acids composition of 10 microalgal species. *Songklanakarin J. Sci. Technol.*, 27, 1179-1187.

Přibyl, P., Eliáš, M., Cepák, V., Lukavský, J. & Kaštánek, P. (2012). Phylogenetic position, ultrastructure, pigment composition and zoosporogenesis of *Trachydiscus minutus* (Eustigmatophyceae, Heterokontophyta.- *J. Phycology* 48: 231-242.

Raja, R. (2008). A Perspective on the Biotechnological Potential of Microalgae. *Crit. Rev. Microbiol.* 34,77–88.

Ramadan, M.F., Asker, M.M.S. & Ibrahim, Z.K. (2008). Functional bioactive compounds and biological activities of *Spirulina platensis* lipids. *Czech J. Food. Sci.*, 26, 211-222.

Richmond, A. (1986). Microalgae of economic potential. In: A.Richmond (Ed.) *Handbook of Microalgal Mass Culture,* pp. 199–243, CRC Press, Inc. Boca Raton, Florida, USA.

Richmond, A. (2004). Handbook of microalgal culture: biotechnology and applied phycology. *Blackwell Sci. Ltd.* UK.

Rodolfi, L., Zitelli, G.Ch., Bassi, N., Padovani, G., Biondi, N., Bonini, G. & Tredici, M.R. (2009). Microalgae for oil: Strain selection, induction of lipid synthesis and outdoor mass cultivation in a low-cost photobioreactor. *Biotechnol. Bioeng.*, 102, 100-112.

Řetovský, R. (1946). Mass culture of some unicellular algae. *Studia Botanica Čechoslovaca* 7, 38-48.

Řezanka, T., Petránková, M., Cepák, V., Přibyl, P., Sigler, K. & Cajthaml, K. (2010). *Trachydiscus minutus*, a new biotechnological source of eicosapentaenoic acid. *Folia Microbiol.*, 55, 265-269.

Řezanka, T., Lukavský, J., Nedbalová, L., & Sigler, K. (2011). Effect of nitrogen and phosphorus starvation on the polyunsaturated triacylglycerol composition, including positional isomer distribution, in the alga *Trachydiscus minutus*. *Phytochemistry* 72: 2342-2351.

Samek, O., Jonáš, A., Pilát, Z., Zemánek, P., Nedbal, L., Tříska, J., Kotas, P. & Trtílek, M. (2010). Raman microspectometry of individual algal cells: Sensing unsaturation of storage lipids in vivo. *Sensors,* 10, 8635-8651.

Sheehan, J., Dunahay, T., Benemann, J. & Roessler, P. (1998). A look back at the U.S. Department of energy's Aquatic Species Program – Biodiesel from algae. Close out report for: U.S. Department of energy's Office of fuels development, pp. 328. *Nat. Renew. Energy Lab.,* Golden, Colorado, USA.

Shen,Y., Yuan, W., Pei, Z., & Mao, E. (2008). Culture of microalga *Botryococcus* in livestock wastewater. Trans. ASABE, 51, 1395-1400.

Shi, S., Valle-Rodríguez, J.O., Siewers, V., & Nielsen, J. (2011). Prospects for microbial biodiesel production. *Biotechnol. J.* 6, 277-285.

Schnepf, E., Niemann, A., & Wilhelm, C. (1996). *Pseudostaurastrum limneticum*, an eustigmatophycean alga with astigmatic zoospores: Morphogenesis, fine structure, pigment composition and taxonomy. *Arch. Protistenkd.,* 146, 237-249.

Staub, R. (1961). Ernährungsphysiologisch autökologische Untersuchungen an der planktonischen Blaualge *Oscillatoria rubescens* DC. Schweiz. *Z. Hydrol.*, 23, 83-198.

Vargas, M.A., Rodriguey, H., Moreno, J., Olivares, H., Del Campo, J.A., Rivas, J. & Guerrero, M.G. (1998). Biochemical composition and fatty acid content of filamentous nitrogen-fixing cyanobacteria. *J. Phycology,* 34, 812-817.

Vazhappilly, R., & Chen, F. (1998). Eicosopentaenoic acid and docosahexaenoic acid production potential of microalgae and their heterotrophic growth. *J. Amer. Oil Chem. Soc.,* 75, 393-397.

Witt, U., Koske, P.H., Kuhlmann, D., Lenz, J., & Nellenn, W. (1981). Production of *Nannochloris* spec. (Chlorophyceae) in large-scale outdoor tanks and its use as food organism in marine aquaculture. *Aquaculture,* 23: 171-181.

Zachleder, V. & Šetlík, I. (1982). Effect of irradiance on the course of RNA synthesis in the cell cycle of *Scenedesmus quadricauda. Biol. Plant.,* 24, 341 – 353.

Zalocar, D.Y, Asselborn, V.M. & Casco, S.L. (1998). Variaciones espaciales y temporales del fitoplancton en un lago subtropical de Argentina. *Rev. Brasil. Biol.,* 58, 359-382.

Review:

By L.Krienitz, Leibnitz – Institute of Freshwater Ecology and Inland Fisheries. Alte Fisherhűtte 2, D - 16775, Stechlin – Neuglobsow, Germany.

In: Algae
Editors: Dagmar Krueger and Helga Meyer
ISBN: 978-1-62081-280-8

Chapter 5

SYSTEMATICS AND TAXONOMIC KEYS FOR THE MARINE GREEN ALGAL FAMILY MONOSTROMATACEAE

Felix Bast*
Centre for Biosciences, Central University of Punjab, Bathinda, Punjab, India

ABSTRACT

Marine green algal family Monostromataceae consists of single cell-layered green seaweeds distributed throughout the world, yet identification of it being one of the most challenging. Monostromatic green seaweeds are all edible and are commercially cultivated in Japan for centuries. A thorough understanding of its systematics is indeed a necessity for any empirical investigation. Working dichotomous taxonomic key for this family is presented for the first time for the ready benefit of field phycologists. Recent advances in the systematics of this family with a special emphasize on the Japanese varieties are also presented.

ALGAE, WHAT ARE THEY?

Algae, which can be roughly defined as a photoautotrophic eukaryote (with the exception of embryophytes) is a ubiquitous and most abundant of the primary producers in the ocean - an area covering 71% of earth. While their biomass constitutes only a meager portion (<1%) the algae are accounted for almost 41% of the world's carbon fixation (Bolin *et al.*, 1977). Despite its overwhelming importance in the biosphere our knowledge of the algal biodiversity remains limited due to several factors such as phycologically unexplored geographical areas, describing new species being time-consuming, and a reduction in the numbers of researchers as well as funds to support algal systematics.

---

* E-mail: felix.bast@gmail.com.

Because of their complex and diverse life cycles, algae have been extensively studied to answer fundamental questions in biology (*e.g.*, *Chlorella*, in discovery of light-independent reactions in photosynthesis [Bassham *et al.*, 1950]), and ecology of terrestrial and aquatic biomes alike. Fossil remains of algae with $CaCO_3$ or silica deposit in their exoskeletons (*e.g.*, diatoms and coccolithophorids) are used routinely for the paleoclimactic reconstructions and thereby to predict climate change. In addition, algae have had a long history in biochemical (*e.g.*, agar-agar from *Gelidium*, carrageenan from *Chondrus*), pharmaceutical (β-carotene from *Dunaliella*, Arachidonic acid from *Parietochloris*) and food (Kombu, Nori and Wakame from seaweeds) industries. In the wake of competing demands during diminishing fossil energy resources, global warming and the world food crisis, algae has attracted much attention lately from the researchers and environmentalists alike as a potential source of renewable energy (the so called "algal biofuel"), as a candidate for carbon capture and sequestration (CCS), and as a future food source.

Classically, algae were classified based on color and distinct morphology (*e.g.*, "red algae lack motile stages"). However, this was found to be insufficient while dealing with the daunting problem of the interrelationships between different groups of eukaryotes. A problem while interpreting phylogeny based on morphology is the difficultly in selecting key character states and assigning phylogenetic values to them, especially in the distantly related groups with substantial morphological differences and of questionable homology. The plasticity of morphological characters resulted in either misclassification or over-classification of recognizable groups that led to a general confusion among phycologists. A significant advancement in the algal systematics took place with the advent of electron microscopy during mid 20th century, when phycologists began to characterize the ultrastructure of algal cells. Since then, the ultrastructural details -especially that of the plastids and the motile cells (flagellar roots and basal bodies) - have been widely taken as a reliable character state in cladistic analysis of various algal taxa because these characteristics were found to be evolutionarily more conserved (Friedl, 1997). By the end of the 20th century, DNA-based molecular systematics had largely superseded ultrastructure–based systematics and it had been shown that the morphological and biochemical diversity of the algae results from their polyphyletic origins within the eukaryotic lineage of the tree of life (Bhattacharya and Medlin, 1995; Stiller and Hall, 1997). The latest among proposed classification schemes for the tree of life, the "six kingdom" model based on molecular, ultrastructural and paleontological evidences (Cavalier-Smith, 2004), groups algal taxa into three kingdoms of the bikonts (ancestrally uniciliate eukaryotes); *viz.*, Protozoa, Chromista and Plantae (Figure 1).

Algae are tremendously diverse; the existence of nine distinct lineages, which are mentioned below, is good evidence for its diversity (Figure 1). Major lineages (divisions) are the Chlorophyta (green algae), Rhodophyta (red algae), Euglenophyta, Chlorarachniophyta, Glaucophyta, Heterokonta, Cryptophyta, Haptophyta, and the dinoflagellates (within the Alveolata). The latter four are grouped together as Chromalveolates or chromophyte algae because they contain various xanthophylls -that make them appear yellow or brown- in addition to the light harvesting pigments chlorophyll *a* and *c*. The algae not only include the world's largest protist giant kelp (*Macrocystis* spp in Heterokonta, that can reach up to 30m in length), but also many unicellular, bacteria- sized coccoid algae (*e.g.*, *Micromonas* spp in Chlorophyta, that have only 1-3 μm cell length).

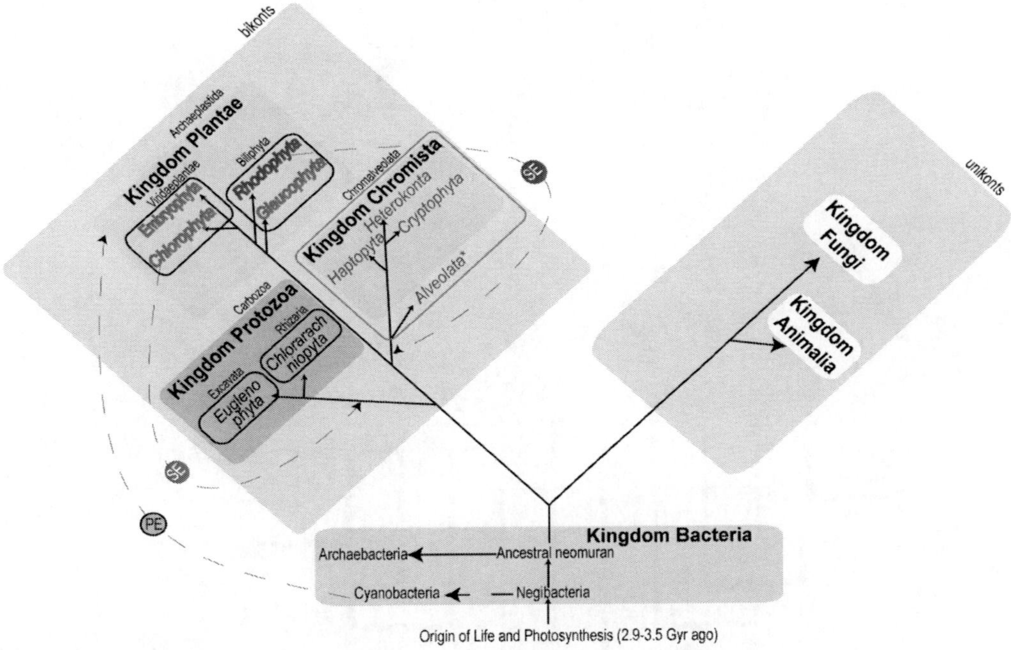

Figure 1. The tree of life based on Cavalier-Smith's "six kingdom" model (redrawn from Cavalier-Smith, 2004 to emphasize kingdoms relevant to algae). Nine algal lineages are indicated in italics. Although a chromalveolate, lineage Alveolata (indicated by an asterisk) belongs to the kingdom Protozoa. PE and SE indicate hypothetical primary and secondary endosymbiotic (endocytobiotic) events respectively, explaining the origin of plastids.

Division Chlorophyta: The Green Algae

Of the nine algal lineages, Chlorophyta is by far the most diverse, with around 16,000 extant species and up to 100,000 species that are estimated to be discovered (Andersen, 1992). Chlorophyta are also found in the greatest diversity of habitats; open oceans, coasts, freshwater, soils, inside shells, on the trees and animals, and even in the lower atmosphere (Bold and Wynne, 1985; Schlichting Jr., 1974). This lineage is estimated to be 1500 million years old (Yoon *et al.*, 2004). The size and structure of the body (thallus) show varying levels of complexity ranging from microscopic motile (e.*g.*, nanoplankton) or nonmotile forms (*e.g.*, lichen phycobionts) to filaments (*e.g.*, *Spirogyra*), bicellular (*e.g.*, *Euastropsis*), giant animal-like colonies (*e.g.*, *Volvox*) and complex levels of tissue differentiation (pseudo-parenchymatous, parenchymatous, or thalloid). Cells can be either uninucleate (*e.g.*, *Monostroma*) or coenocytic (*e.g.*, *Caulerpa* and *Bryopsis*).

Division Chlorophyta (Pascher, 1914) is characterized by the presence of double membrane-bound plastids containing chlorophylls *a* and *b*, accessory pigments β-carotene and xanthophylls (as in bryophytes) and a unique stellate structure linking 9 pairs of microtubules in the basal body of flagella, when present (Mattox and Stewart, 1984; Sluiman, 1985; Bremer *et al.*, 1987; Kenrick and Crane, 1997). Flagella lack tripartite tubular hairs (Moestrup, 1982; Mattox and Stewart, 1984). Main storage product is starch (amylose and amylopectin) and is stored inside plastids as grains (Dodge, 1973). Cell walls, when present, are usually composed of cellulose (Graham and Wilcox, 2000).

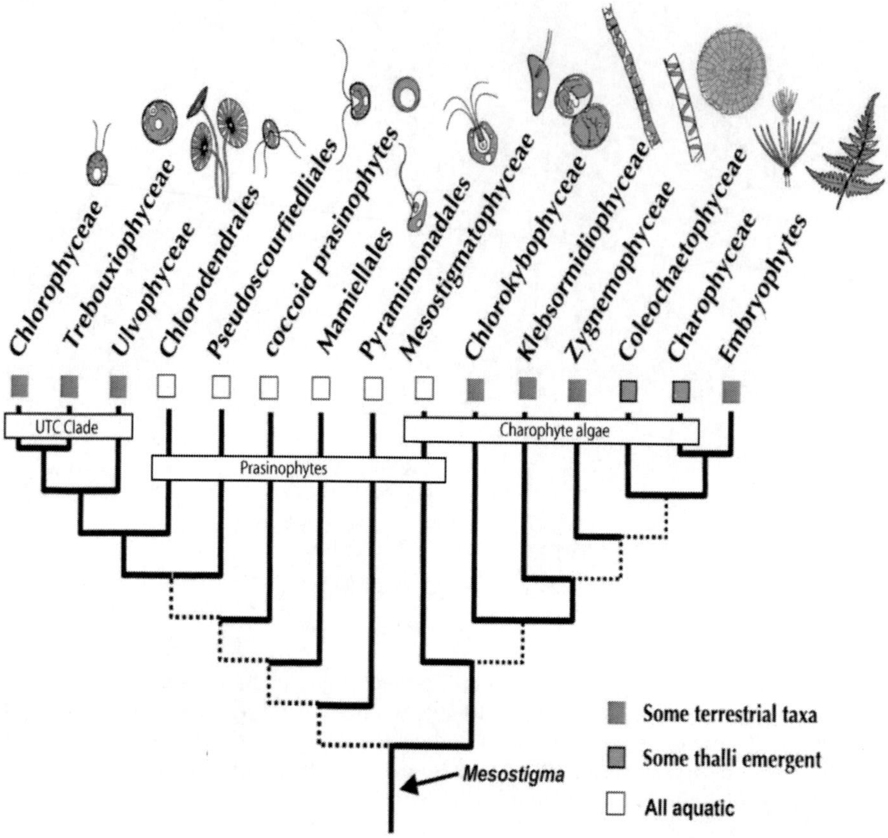

Figure 2. Internal lineages of the green algae inferred from DNA sequence data (redrawn from Lewis and McCourt, 2004).

Green algae have evolved into three major clades; charophyte algae, prasinophytes and "UTC clade" (Figure 2). Some taxonomists elevate the charophyte clade to a rank of division, *i.e.*, Charophyta (Cavelier-Smith, 2004), while the others consider this clade as a class within Chlorophyta. Charophytes, although relatively poor in species diversity, are believed to be the ancestral group of the Embryophytes-a clade consisting of over 100,000 species (Pickett-Heaps, 1969). Charophytes show great diversity in cellular organization, from flagellate unicells and filaments (branched or unbranched) to fairly complex forms, sometimes referred to as parenchymatous. Prasinophytes, also known as "ancestral green flagellate" (AGF), are widely regarded to be the most primitive green algae (Mattox and Stewart, 1984). Oldest fossils of green algae ever discovered are phycoma stage of prasinophytes, dated 1.2 bya (Tappan, 1980). Some of the best-known bloom-forming marine planktonic algae are prasinophytes (O'Kelly *et al.*, 2003). Prasinophytes are currently considered to be representatives of early diverging clades of uncertain taxonomic affinity (Fawley *et al.*, 2000). The name of the third clade, "UTC", stands for the three groups it contains; *viz.*, Ulvophyte, Trebouxiophyte and Chlorophyte groups. Vast majority of the described species of green algae are part of this triad. Monophyly of UTC clade has been strongly supported by various molecular studies based on nrDNA 18S (nucleoribosomal DNA gene for ribosomal small subunit) and has indicated that the Ulvophyte group is sister to the other two groups (Mishler *et al.*, 1994; Friedl, 1995; Krienitz *et al.*, 2001). These three groups are generally considered to be three distinct classes of green algae, *viz.*, Ulvophyceae, Trebouxiophyceae and Chlorophyceae (Mattox and Stewart, 1984).

Of the five classes of green algae, only Ulvophyceae is predominantly marine and it contains almost all known marine green macroalgae (the so-called green seaweeds). Ulvophyceae (Mattox and Stewart, 1984) is one of the algal classes that are exclusively classified based on ultrastructural characteristics. Members of this class are characterized by the presence of one or two pairs of flagella without mastegonemes in the motile cells, basal bodies having four microtubular rootlets arranged in stellar pattern, cell division by furrowing with centric, persistent and closed mitotic spindle fibers, and diplobiontic life cycle in free-living forms (O'Kelly and Floyd, 1984; Bold and Wynne, 1985). Various forms of thallus are present in Ulvophyceans; *e.g.*, filamentous (branched or unbranched), membranous (distromatic or monostromatic) or tubular. Cells can be either uninucleate or coenocytic. Some taxa have heavily calcified thalli (*e.g.*, *Halimeda*) and some have abundant deposits of orange pigment that make them appear brown or orange (*e.g.*, *Trentepohlia*). So far, molecular data (nrDNA 18S and *rbcL*- the gene encoding large subunit of RuBisCO-ribulose-1,5-bisphosphate carboxylase/ oxygenase) has not been able to provide solid support for the recognition of a monophyly of the class Ulvophyceae *sensu* Mattox and Stewart (Zechman *et al.*, 1990; Watanabe *et al.*, 2001; López-Bautista and Chapman, 2003; Lewis and McCourt, 2004). Within this class, there are five groups (orders) recognized by many contemporary classification schemes, *viz.*, Ulotrichales, Ulvales, Bryopsidales (Caulerpales) and Dasycladales, Cladophorales (including Siphonocladales), and Trentepohliales (Guiry and Guiry, 2008; Index Nominum Algarum, 2009), as illustrated in Figure 3.

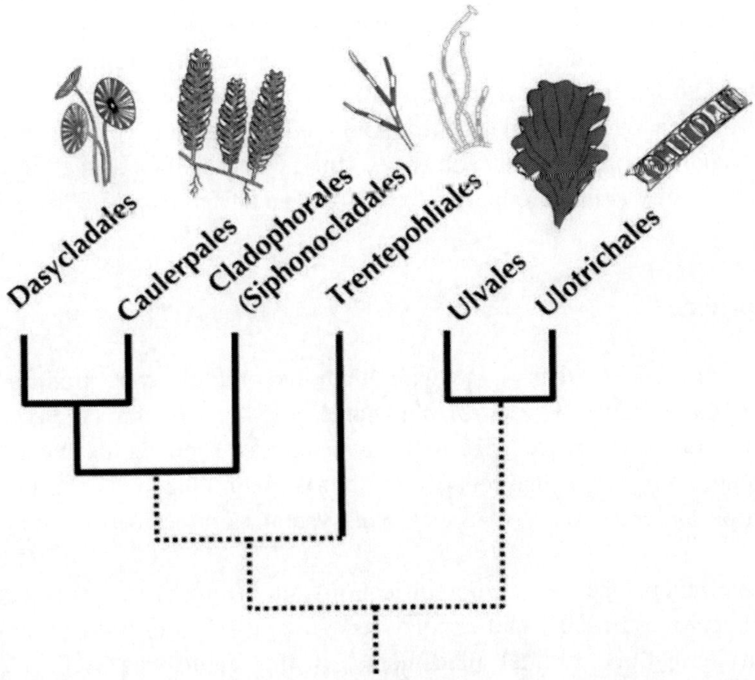

Figure 3. Internal lineages of Ulvophyceae based on DNA sequences (adapted from Lewis and McCourt, 2004).

Orders Ulotrichales (Borzi, 1895) and Ulvales (Blackman and Tansley, 1902) contain many green seaweed genera commonly seen across the world. Traditionally, Ulotrichales included only unbranched filaments and Ulvales included branched filaments or blades, but as ultrastructural and molecular data became available, several Chlorophycean taxa were added to either of these orders. Morphological forms in Ulotrichales include branched filaments (as in *Urospora*), unbranched filaments (as in *Ulothrix*) and membranous monostromatic blades (*Monostroma*). Most of the Ulotrichalean species have heteromorphic diplohaplontic alternation of generation with typical microscopic sporophytes known as *Codiolum*. Based on the presence of *Codiolum* stage, Kornmann (1973) erected a new class within Chlorophyta, the Codiolophyceae. However, the International Code of Botanical Nomenclature (ICBN) rejected this class because no Latin diagnosis was provided. Mattox and Stewart (1984) did not recognize this class as well because order Ulvales was not included. Genus named *Codiolum*, characterized based on aforementioned *Codiolum* stage, was later found to be sporophytic stage in the life cycle of *Urospora*, a later synonym. Taxonomic situation of Ulotrichales is problematic, especially because circumscription of this order, and its grouping under class Ulvophyceae was based on ultrastructural studies conducted in just one species, *Ulothrix zonata* (O'Kelly and Floyd, 1984). Ultrastructural characteristics of the type species of *Ulothrix*, *U. tenuissima* Kützing is unknown and because ICBN established that order name should be connected to the name of type genus and its type species; taxonomic validity of this order is still uncertain. On the other hand, members of the Ulvales have isomorphic alternation of generations. Morphologic switch between tubular (as in *Enteromorpha*) and membranous distromatic blade (as in *Ulva*) forms of Ulvales have been suggested to have evolved multiple times (Tan *et al.*, 1999). Molecular work indicated that the genera *Ulva* and *Enteromorpha*, two of the familiar Ulvacean seaweeds worldwide, are monophyletic (Hayden and Waaland, 2002; Hayden *et al.*, 2003). Several systematic studies argue in favor of merging Ulvales into Ulotrichales (*e.g.,* Papenfuss, 1960). Molecular phylogenetic studies in general show that these two orders are monophyletic, although bootstrap support for Ulotrichales is weak or absent (Watanabe *et al.*, 2001; Friedl and O'Kelly, 2002; Hayden and Waaland, 2002; O'Kelly *et al.*, 2004a, 2004b, 2004c; Lindstrom *et al.*, 2006).

Monostromataceae

All single cell-layered green algae with blade-like thallus were traditionally grouped under the eponymous genus *Monostroma* (Thuret, 1854). Kunieda (1934) erected family Monostromaceae (later synonym for Monostromataceae Kunieda ex Suneson, 1947) to include this genus. Various and often contradictory taxonomic revisions have resulted in over-classification as well as lack of a clear-cut systematic placement of this Ulvophycean group.

For example, there are two lectotypifications in *Monostroma*; first recognizing *M. oxyspermum* (Papenfuss, 1960), and second recognizing *M. bullosum* (Kornmann, 1964). There were at least three ordinal placements of this family as well, *viz.*, Codiolales (Kornmann, 1964; van den Hoek *et al.*, 1995), Ulotrichales (Gayral, 1964; Chapman and Chapman, 1973; Floyd and O'Kelly, 1990; Gabrielson *et al.*, 2000; Graham and Wilcox, 2000) and Ulvales (Bliding, 1968; Vinogradova, 1974; Bold and Wynne, 1985).

Systematic history of Monostromataceae has been reviewed in great detail by Tatewaki (1969) and O'Kelly *et al.* (1984). In summary, the following are the major revisions. Gayral (1964) grouped *M. angicava, M. grevillei* and *M. bullosum* under *Ulvopsis* due to shared ontogeny (Disc-Sac-Blade; DSB) and swarmer release (simultaneously through an irregular rent), and reserved genus *Monostroma* only for asexual species with typical ontogeny (presence of a filament stage) and zoid release mechanism (*en-masse* without pore). Inversely, Kornmann (1964) and Bliding (1968) proposed to remove asexual members from *Monostroma*. Vinogradova (1969) erected two monotypic genera *Protomonostroma* and *Gayralia,* to accommodate the asexual members *M. undulatum* and *M. oxyspermum,* respectively. *Monostroma fuscum* and *M. obscurum* have been reclassified with resurrected genus *Ulvaria* in order Ulvaceae because of the isomorphic life cycle pattern (Gayral, 1964). *Monostroma leptodermum* and *M. zostericola* have been grouped under the new genus *Kornmannia* due to shared life cycle and ontogenetic patterns and placed under Ulvales due to typical flagellate release mechanism in which swarmers are released one by one through a gametangial exit pore (Bliding, 1968).

Monostroma groenlandicus has been included in the genus *Capsosiphon* due to similarity in habit (cylindrical gametophyte), thallus ontogeny (filament-tube) and swarmer release mechanism (*en-masse*, enclosed within hyaline sheath; Vinogradova, 1969).

Taxonomic confusion has largely been caused by a lack of synapomorphic character due to polyphyly in this group of algae. Many studies have concluded that Ulvophycean algae show considerable phenotypic plasticity and therefore diagnostic character such as "monostromatic blade" is not taxonomically reliable.

For example, green algae belonging to *Prasiola* have macroscopic monostromatic thalli that closely resemble *Kornmannia*, however they belong to an entirely different class (Trebouxiophyceae). A green-tide forming single cell-layered algae that superficially resembles *Monostroma* isolated from west coast of Finland turned out to be a morphotype of tubular *Ulva* (Blomster *et al.*, 2002). Abiotic factors such as nutrient supply (Valiela *et al.*, 1997) and salinity (Reed and Russsel, 1978) are believed to be inducing morphological changes in green algae.

Phenotypical polymorphism induced by biotic factors is known in green algae for a long time. Bonneau (1977) discovered that adding extracts containing marine bacteria isolated from thallus of *Ulva* to axenic cultures of the same algae resulted in a change of its morphology from blade to tube. Recently, specific bacterial strains isolated from *M. oxyspermum* had been demonstrated to induce morphogenetic changes in the axenic cultures of this alga, as well as *Ulva pertrusa* and *Ulva intestinalis*. Through phylogenetic analyses, these strains had been identified as belonging to *Cytophaga-Flavobacterium-Bacteroides* (CFB) complex (Matsuo *et al.*, 2003).

Further investigations on this bacterium lead to biochemical characterization of a morphogenetic inducer (Thallusin), and confirmation of its importance for the natural growth of these algae (Matsuo *et al.*, 2005).

A diagnostic dichotomous key to aid in the identification of various monostromatic green algal species is presented in Table 1. As one can observe, field identification of these algae is extremely challenging and is limited only to a few species; the rest demands life cycle and other culture experiments that are often time-consuming.

Table 1. Diagnostic dichotomous key for the monostromatic green algae[1,2]

1. Fronds macroscopic, tubular, wall of tube one cell thick in cross section : **2**
1. Fronds macroscopic, bladelike, monostromatic in cross section : **4**
2. No rhizoidal filaments at the base of tube; Tube arising from parenchymatous cushion; Eulittoral to supralittoral. ***Blidingia*** (in part)
2. Rhizoidal filaments at the base of tube; Tube arising initially from uniseriate filament; Sublittoral, Eulittoral or supralittoral : **3**
3. Fronds are hollow except at base and ends of branches; pale to bright green. ***Ulva*** (in part)
3. Fronds are often filled with gelatinous substance and rarely hollow; Fronds very slender at base and broadened towards apex, up to 1mm in diameter and 5-10 cm in height; Yellowish or brownish green; Estuarine supralittoral. ***Capsosiphon groenlandicus***
4. Blade typically less than 5 cm in height, cells typically less than 8 μm in diameter : **5**
4. Blade typically larger than 5 cm in height, cells typically larger than 8 μm in diameter : **7**
5. Blades typically less than 2 cm in height, narrowing to short stipiate region at base; Epilithic in supralittoral zone; Zoids released from fronds are nonphototactic, biflagellate . ***Prasiola*** (in part)
5. Blades typically 3-5 cm with ragged outline; Often epiphytic on seagrasses, *Fucus* or *Halosaccion*; Zoids released from fronds are nonphototactic, quadriflagellate; Marine eulittoral/sublittoral; Spring ephemeral; DSB ontogeny : **6**
6. With tubular stipe, Cells are 6-8 μm in diameter. ***Kornmannia leptoderma***
6. Without tubular stipe, Cells are 2.5-5 μm in diameter. ***Kornmannia zostericola***
7. Limnetic; Plants usually monoecious; Gametes are biflagellate, phototactic; Heteromorphic alternation; Shell-boring *Codiolum*-sporophyte; DSB Ontogeny. ***Monostroma bullosum***
7. Marine or Estuarine : **8**
8. Zoids released from the fronds are nonphototactic, quadriflagellate; Zoids exhibit characteristic clumping behavior; Marine eulittoral/sublittoral; Spring ephemeral; Dimorphic asexual life cycle with *Codiolum*-sporophyte (sexual life cycle not reported); *En masse* zoid release by wall dissolution; Zoidangia disintegrates upon zoid release; FB ontogeny. ***Protomonostroma undulatum***
8. Zoids released from the fronds are phototactic, biflagellate : **9**
9. Life cycle is isomorphic alternation (asexual life cycle not reported); Dark or bright green; Turns to olive green or black upon drying; Singular zoid release through pore; Zoidangia remain on blade after zoid release; FSB ontogeny. ***Ulvaria obscura***
9. Life cycle is either heteromorphic alternation or monomorphic asexual: **10**
10. Prostrate disc stage in the ontogeny (DSB); *En masse* zoid release by irregular rent; Zoidangia remain in blade after zoid release : **11**
10. Erect filament stage in the ontogeny (FB or FSB); *En masse* zoid release by wall dissolution; Zoidangia disintegrates upon zoid release : **12**
11. Life cycle is monomorphic asexual (sexual life cycle not reported); Marine; Reported only from temperate to polar regions. ***Monostroma arcticum****
**M. antarticum* may be a related species but life cycle information is not known
11. Life cycle is dimorphic alternation (asexual life cycle not reported); Marine eulittoral/ sublittoral; Shell-boring *Codiolum*-sporophyte: ***Monostroma grevillei**** or ***M. angicava****
*Discerning between these two species is difficult or impossible, may be synonymous.
12. FB Ontogeny; Spring ephemeral; Marine or estuarine eulittoral; Life cycle is sexual or asexual; If sexual, dimorphic alternation; Non shell-boring *Codiolum*-sporophyte. ***Monostroma kuroshiensis***
12. FSB ontogeny : **13**
13. Present year round; Typically estuarine eulittoral; Life cycle is asexual (sexual lifecycle not reported); Thallus often sac shaped. ***Monostroma oxyspermum***
13. Spring ephemeral; Life cycle is dimorphic alternation (asexual life cycle not reported); Non shell-boring *Codiolum*-sporophyte; Marine or estuarine eulittoral. ***Monostroma nitidum***

[1] Information based on following references: Gayral, 1964; Bliding, 1968; Tatewaki, 1969; Gabrielson *et al.*, 2000; Guiry and Guiry, 2008.

[2] Expansion of abbreviations: DSB: Disk-Sac-Blade; FSB:Filament-Sac-Blade; FB:Filament-Blade.

MONOSTROMA KUROSHIENSIS-NITIDUM COMPLEX

Monostroma kuroshiensis (Yendo) Bast *et al.* and *M. nitidum* share a lot of taxonomic features in common including life cycle, basic thallus ontogeny pattern and gamete release mechanism. Both have heteromorphic diplohaplontic life cycle with leafy gametophytes and *Codiolum*-sporophytes. Gametophyte releases positively phototactic, biflagellate gametes by thallus disintegration. Sexual fusion is anisogamic and subsequently, zygotes become negatively phototactic. Successive divisions and enlargement of the zygote produces a microscopic *Codiolum*-sporophyte. *Codiolum*-sporophyte releases quadriflagellate zoospores that germinate to form either of the gametophytes. Systematically, these two species were distinguished based only upon the frond thickness; *i.e.*, "thicker" species being *M. nitidum* (Wittrock, 1866). There have been several reports on the thallus ontogeny of *M. kuroshiensis*, which is of FB (Filament-Blade) pattern. While this ontogenetic pattern can be considered as a typical characteristic of this complex, there has been one report (Arasaki, 1949) that mentions FSB (Filament-Blade-Sac) ontogeny for *M. nitidum*. Although the differences between them warrant further investigation, we tentatively call these two very similar algae as *Monostroma kuroshiensis-nitidum* complex, or *K-N*.

Commonly found in the eulittoral zones of marine and estuarine environments, algae belonging to this complex are crucial players contributing to the ecology of the coastal ecosystem. Along with other abundant primary producers, *K-N* contributes to the growth and survival of the whole niche of coastal biota. Many small benthic invertebrates, collectively known as meiofauna, utilize the assemblage of *K-N* as a shelter as well. Therefore, a good understanding of the ecophysiology as well as biogeography of this complex has become increasingly important in our overall comprehension of the coastal ecosystem.

Algae belonging to this complex have been used for human consumption in East Asia- especially in Japan- since time immemorial. At present, the *K-N* complex is the most important commercial green algae in Japan, constituting about 90% of the total green algal cultivation (Nisizawa *et al.*, 1987). In and around Ise Bay, Mie Prefecture, Central Japan, is the epicenter of *K-N* cultivation, with a vast majority of the edible monostromatic seaweed, known in Japanese as "*hitoegusa*" coming from this region (Figure 4). Seeding method used in the cultivation can either be natural (*e.g.*, Ise Bay) or artificial (*e.g.*, Shimanto Estuary, Kochi Prefecture). In the natural seeding method, plants derived from the naturally deposited zoospores on the culture nets are harvested. On the other hand, in the artificial seeding method a large quantity of zygotes –as obtained from *in vitro* fertilization of isolated gametes by the end of growth period- are grown throughout the summer. The resulting *Codiolum*-sporophytes are then treated with high intensity of light to induce zoospore release. Culture nets are immersed in the concentrated zoospore solution under dark conditions to facilitate successful attachment of released zoospores on the nets. These "seeded" nets are subsequently installed in the attached fabrication of wooden sticks in the coastal waters and the height of nets are adjusted such as to provide adequate immerse-in and drying-out effects with each tidal range. Upon reaching the highest size, thalli are harvested and processed. Harvested thallus is boiled down in soy sauce to make a jam-like product, *tsukudani-nori* while dried sheets (*hoshi-nori*) are used as sushi wraps. Other popular products are roasted (*yaki-nori*) and seasoned (*ajitsuke-nori*) *Monostroma*.

Figure 4. *Monostroma kuroshiensis* from Matsusaka, Mie Prefecture, Japan. a. Fronds growing luxuriously on the intertidal cultivation nets. b. A close-up view on the cultivation net. c. *In vitro* phototactic isolation of gametes using a capillary tube by the first author.

Several independent pharmacological studies suggest that the algae belonging to this complex are potential pharmaceutical agents. Rhamnan sulfate, a water-soluble sulfated polysaccharide isolated from the fronds of *K-N*, has shown to possess antiviral activity against pathogenic viruses such as Human Immunodeficiency Virus type 1 (HIV-1), herpes simplex virus type 1 (HSV-1) and cytomegalovirus (HCMV) *in vitro* (Lee *et al.*, 1999). Rhamnan sulfate and other sulfated polysaccharides isolated from this complex also have been demonstrated to possess antithrombin activity (Lee *et al.*, 1998; Zhang *et al.*, 2008). Of several betains isolated from *K-N*, β-homobetain has been shown to lower plasma cholesterol level in rats (Abe and Kaneda, 1973). Hot water extract of the *K-N* had been demonstrated to increase longevity of the animals implanted with leukemia cells (Yamamoto *et al.*, 1982). There has also been a successful attempt on the isolation of protoplasts from the vegetative thalli of this complex, for a potential use as seed stock (Chen, 2002).

This ephemeral species had been reported only during the colder part of the year (Kida, 1990). Typically, growth season starts in late autumn and ends in mid spring. The *K-N* complex is observed only along the Central and SW Japanese coasts influenced by the warm water Kuroshio Current. Geographical history of Japan might have had an influence on the distribution pattern of *K-N* complex. After the recession of the last ice age around 10,000 years ago, tremendous rise in sea level resulted in the isolation of the four main islands (*i.e.*, Hokkaido, Honshu, Shikoku, and Kyushu) from the Korean peninsula (Ohta and Yonekura, 1987). Seawater from the Pacific infiltrated to the Sea of Japan and Seto Inland Sea, both of which had been freshwater lakes during the ice age. Re-colonization of the marine algae from Southern Pacific might have taken place during this period along the warmer waters of the Southern and Western Japan, accounted by Kuroshio and Tsushima currents, respectively.

By a novel, polyphasic strategy that involved classical culture-based methods such as mating tests and microscopic investigations of life cycle and ontogeny, as well as modern genome-based techniques such as DNA barcoding, mRNA secondary structure analysis and multi-local phylogeny reconstruction, taxonomic reappraisal of this genus has recently been proposed (Bast, 2010). Natural and cultivated sexually reproducing environmental samples, as well as sexually reproducing and serendipitously discovered asexually reproducing ecotypes of monostromatic green alga from Tosa Bay, Japan were discovered to be conspecific and was identified as *Monostroma kuroshiensis* (Yendo) Bast *et al* (Bast et al, 2009a). Patterns of seasonal fluctuations in its thallus lengths were habitat specific and recur annually. Both appearance and decay of thalli were earlier at high saline habitats, suggesting that salinity positively influences either maturation of sporophytes or senescence of gametophytes (Bast *et al*, 2009b). Gametogenesis in *M. kuroshiensis* occurs in discontinuous patches along the frondal apex and the gametes release synchronously in a posterior faced linear fashion by the dehiscence of gametangial sheath, leading to the thallic disintegration (Bast and Okuda, 2010). The overall primary sex ratio of this alga was about 1:1 which is likely to reflect Fisherian selection (Bast *et al*, 2009c). Results from life cycle, thallus ontogeny, gametangial ontogeny and phylogenetic analyses suggest that this alga is evolutionarily more close to *M. oxyspermum* than other green algae. As there are conspecific members having different life cycles, type of life cycle is not a valid diagnostic character for the species circumscription in Monostromataceae. Life cycle may be an adaptive strategy to increase fitness of this algae during environmental stress. Biogeographic isolates of *Monostroma* sp. distributed along Kuroshio Coast of Japan were panmictic and had similar sequences for the conserved nucleoribosomal DNA (nrDNA) 18S gene, however, a substantial variance was observed in the highly variable first internal transcribed spacer (ITS1) sequences that are suggestive of an emerging sympatric speciation (Bast, 2010).

Whereas differences in morphology, color and habitat of various monostromatic algae remains so feeble, culture-based investigation on its life cycle and thallus ontogeny remains the only method to conclusively identify the members of monostromataceae. Hopefully, recently developed ITS based DNA barcoding (Bast, 2010) will bring about the cryptic monostromatic marine biodiversity. As we expand our search to more and more phycologically unexplored sea shores, the unambiguous green algal biodiversity will keep on unfolding, for which it is expected that the taxonomic key of monostromatic green algae presented herein will be of some help.

References

Abe S & Kaneda T (1973) Studies on the effect of marine products on cholesterol metabolism in rats-VIII. The isolation of hypocholesterolemic substance from green algae. *Bull. Jpn. Soc. Sci. Fish.* 39:383-389.

Andersen RA (1992) Diversity of eukaryotic algae. *Biodivers. Conserv.* 1:267-292.

Arasaki S (1949) On the *Monostroma* found in Ise and Kikawa Bay. *Nippon Suisan Gakkaishi* 15(3):137-143.

Bassham J, Benson A. & Calvim M (1950) The path of carbon in photosynthesis. *J. Biol. Chem.* 185:781-787.

Bast F, Shimada S, Hiraoka M & Okuda K (2009a) Asexual life history by biflagellate zoids in *Monostroma latissimum* (Ulotrichales). *Aquat. Bot.* 91:213-218.

Bast F, Shimada S, Hiraoka M & Okuda K (2009b) Seasonality and thallus ontogeny of edible seaweed *Monostroma latissimum* (Kützing) Wittrock (Chlorophyta Monostromataceae) from Tosa Bay Kochi Japan. *Hydrobiologia* 630:161-167.

Bast F, Hiraoka M & Okuda K (2009c) Spatiotemporal Sex Ratios of a Dioecious Marine Green Alga: *Monostroma latissimum* (Kützing) Wittrock. *International Journal on Algae* 11(2):141-150.

Bast F & Okuda K (2010) Gametangial Ontogeny in Intertidal Green Alga: *Monostroma latissimum* (Kützing) Wittrock. *International Journal of Plant Reproductive Biology* 2 (1): 1-5.

Bast F (2010) Comparative Ecophysiology and Phylogeography of Monostroma in Southern Japan. Ph.D. Thesis, Kochi University, Kochi, Japan.

Bhattacharya D & Medlin L (1995) The phylogeny of plastids: a review based on comparisons of small-subunit ribosomal RNA coding regions. *J. Phycol.* 31:489-498.

Blackman FF & Tansley AG (1902) A revision of the classification of the green algae. *New Phytologist* 1:17-24, 47-48, 67-72, 89-96, 114-120, 133-144, 163-168, 189-192, 213-220, 238-244.

Bliding C (1963) A critical survey of European taxa in *Ulvales*, Part I: *Capsosiphon, Percursaria, Blidingia, Enteromorpha. Opera Bot. Soc. Bot. Lund, Sweden.* 8 (3):1-160.

Bliding C (1968) A critical survey of European taxa in Ulvales Part II: *Ulva, Ulvaria, Monostroma, Kornmannia. Bot. Notiser.* 121:535-629.

Blomster J, Back S, Fewer DP, Kiirikki M, Lehvo A, Maggs CA & Stanhope MJ (2002) Novel morphology of *Enteromorpha* forming green tides. *Am. J. Bot.* 89:1756-1763.

Bold HC & Wynne MJ (1985) *Introduction to the Algae*. 2nd Edition. Prentice-Hall, Inc., Englewood Cliffs, New Jersey, USA. 720 pp.

Bolin B, Degens ET, Duvigneaud P & Kempe S (1977) The global biogeochemical carbon cycle. *In* Bolin B, Degens ET, Kempe S & Ketner P [Eds.], *The global carbon cycle.* J. Wiley & Sons Inc, New York, USA. pp 1-53.

Bonneau ER (1977) Polymorphic behavior of *Ulva lactuca* (Chlorophyta) in axenic culture. 1. Occurrence of *Enteromorpha*-like plants in haploid clones. *J. Phycol.* 13:133-140.

Borzì A (1895) Studi algologici. Saggio di richerche sulla biologia delle Alghe. Fasc. 2:119-378.

Bremer K, Humphries CJ, Mishler BD & Churchill SP (1987) On cladistic relationships in green plants. *Taxon* 36:339-349.

Cavalier-Smith, T (2004) Only six kingdoms of life. *Proc. R. Soc. Lond.* 271:1252-1262.

Chapman VJ & Chapman DJ (1973) *The Algae.* 2nd Ed., MacMillan, London, UK. 497 pp.

Chen Y-C (2002) Development of protoplasts from holdfasts and vegetative thalli of *Monostroma latissimum* (Chlorophyta, Monostromatacae) for algal seed stock. *J. Phycol.* 34:1075-1081.

Dodge JD (1973) *The Fine Structure of Algal Cells.* Academic Press. New York, USA. 261 pp.

Fawley MW, Yun Y & Qin M (2000) Phylogenetic analyses of 18s rDNA sequences reveal a new coccoid lineage of the Prasinophyceae (Chlorophyta). *J. Phycol.* 36:387-393.

Floyd GL & O'Kelly CJ (1990) Phylum Chloropyta Class Ulvophyceae. *In* Margulis L, Corliss JO, Melkonian M& Chapman DJ [Eds.], *Handbook of Protoctista*. Jones and Bartlett, Boston, USA. pp. 617-635.

Friedl T (1995) Inferring taxonomic positions and testing genus level assignments in coccoid green lichen algae-a phylogenetic analysis of 18S ribosomal-RNA sequences from *Dictyochloropsis reticulata* and from members of the genus *Myrmecia* (Chlorophyta, Trebouxiophyceae Cl. Nov). *J. Phycol.* 31:632-639.

Friedl T (1997) The evolution of the green algae. *In* Bhattacharya D [Eds.], *Origins of Algae and Their Plastids*. Springer-Verlag, Wien, Germany. 287 pp.

Friedl T & O'Kelly CJ (2002) Phylogenetic relationships of green algae assigned to the genus *Planophila* (Chlorophyta): evidence from 18S rDNA sequence data and ultrastructure. *Eur. J. Phycol.* 37:373-384.

Gabrielson PW, Widdowson TB, Lindstrom SC, Hawkes MJ & Scagel RF (2000) *Keys to the Benthic Marine Algae and Seagrasses of British Columbia, southeast Alaska, Washington and Oregon*. Department of Botany, University of British Columbia, Vancouver, Canada. 209 pp.

Gayral P (1964) Sur le démembrement de l'actual genre *Monostroma* Thuret (Chlorophycées, Ulotrichales s.l.). *C. R. Acad. Sci. Paris* 258:2149-2152.

Graham LE & Wilcox LW (2000) *Algae*. Prentice Hall, Upper Saddle River, NJ, USA.

Guiry MD & Guiry GM (2008) AlgaeBase. World-wide electronic publication, National University of Ireland, Galway. Available at: http://www.algaebase.org [accessed June 13th 2011].

Hayden HS & Waaland JR (2002) Phylogenetic systematics of the ulvaceae (Ulvales Ulvophyceae) using chloroplast and nuclear DNA sequences. *J. Phycol.* 38:1200-1212.

Hayden HS, Blomster J, Maggs CA, Silva PC, Stanhope MJ & Waaland JR (2003) Linnaeus was right all along: *Ulva* and *Enteromorpha* are not distinct genera. *Eur. J. Phycol.* 38:277-294.

Index Nominum Algarum (2009) World-wide electronic publication, University Herbarium, University of California, Berkeley. Compiled by Paul Silva. Available at: http://ucjeps.herb.berkeley.edu/INA.html [accessed 3rd June 2011].

Kenrick P & Crane PR (1997) The origin and early evolution of plants on land. *Nature* 389:33-39.

Kida W (1990) Culture of seaweeds *Monostroma*. *Mar. Behav. Physiol.* 16:109-131.

Kornmann P (1964) *Über Monostroma bullosum* (Roth) Thuret und *M. oxyspermum* (Kützing) Doty. *Helgol. Wiss. Meeresuntersuch.* 11:13-21 (in German).

Kornmann P (1973) Codiolophyceae, a new class of Chlorophyta. Helgoländer Wissenschaftliche *Helgol Wiss Meeresuntersuch*. 25:1-13.

Krienitz L, Ustinova I, Friedl T & Huss VAR (2001) Traditional generic concepts versus 18S rRNA gene phylogeny in the green algal family Selenastraceae (Chlorophyceae, Chlorophyta). *J. Phycol.* 37:852-865.

Kunieda H (1934) On the life history of *Monostroma*. *Proceed. Imp. Acad. Tokyo* 10(2):103-106.

Lee J-B, Yamagaki T, Maeda M & Nakanishi H (1998) Rhamnan sulfate from cell walls of *Monostroma latissimum*. *Phytochemistry* 48:921-925.

Lee J-B, Hayashi K, Hayashi T, Sankawa U & Maeda M (1999) Antiviral activities against HSV-1, HCMV, and HIV-1 of Rhamnan sulfate from *Monostroma latissimum*. *Planta Med.* 65:439-441.

Lewis LA & McCourt RM (2004) Green algae and the origin of land plants. *Am. J. Bot.* 91:1535-1556.

Lindstrom CS, Hanic LA & Golden L (2006) Studies of the green alga *Percursaria dawsonii* (=*Blidingia dawsonii* comb. nov, Kornmanniaceae, Ulvales) in British Columbia. *Phycol. Res.* 54:40-56.

López-Bautista JM & Chapman RL (2003) Phylogenetic affinities of the Trentepohliales inferred from small-subunit rDNA. *Int. J. Syst. Evol. Microbiol.* 53:2099-2106.

Matsuo Y, Suzuki M, Kasai H, Shizuri Y & Harayama S (2003) Isolation and phylogenetic characterization of bacteria capable of inducing differentiation in the green alga *Monostroma oxyspermum*. *Environmental Microbiology* 5:25-35.

Matsuo Y, Imagawa H, Nishizawa M & Shizuri Y (2005) Isolation of an algal morphogenesis inducer from a marine bacterium. *Science* 307:1598.

Mattox KR & Stewart KD (1984) Classification of the green algae: a concept based on comparative cytology. *In* Irvine DEG & John DM [Eds.], *The Systematics of Green Algae*. The Systematics Association Special Volume 27, Academic Press, London, UK. pp. 29-72.

Mishler B, Lewis DIA, Buchheim MA, Renzaglia KS, Garbary DJ, Delwiche CF, Zechman FW, Kantz TS & Chapman RL (1994) Phylogenetic relationships of the ''green algae'' and ''bryophytes.'' *Ann. Mo. Bot. Gard.* 81:451-483.

Moestrup O (1982) Flagellar structure in algae: a review, with observations particularly on the Chrysophyceae, Phaeophyceae (Fucophyceae), Euglenophyceae, and Reckertia. *Phycologia* 21:427-528.

Nisizawa K, Noda H, Kikuchi R & Watanabe T (1987) The main seaweed foods in Japan. *Hydrobiologia* 151/152:5-29.

O'Kelly CJ & Floyd GL (1984) Correlations among patterns of sporangial structure and development, life histories, and ultrastructural features in the Ulvophyceae. *In* Irvine DEG & John DM [Eds.], *The Systematics of Green Algae*, The Systematics Association Special Volume 27, Academic Press, London, UK. pp. 121-156.

O'Kelly CJ, Floyd GL & Dube MA (1984) The fine structure of motile cells in the genera *Ulvaria* and *Monostroma* with special reference to the taxonomic position of *Monostroma oxyspermum* (Ulvophyceae, Chlorophyta). *Pl. Syst. Evol.* 144:179-199.

O'Kelly CJ, Sieracki ME, Their EC & Hobson IC (2003) A transient bloom of *Ostreococcus* (Chlorophyta, Prasinophyceae) in West Neck Bay, Long Island, New York. *J. Phycol.* 39:850-854.

O'Kelly CJ, Bellows WK & Wysor B (2004a) Phylogenetic position of *Bolbocoleon piliferum* (Ulvophyceae, Chlorophyta): evidence from reproduction, zoospore and gamete ultrastructure, and small subunit rRNA gene sequences. *J. Phycol.* 40:209-222.

O'Kelly CJ, Wysor B & Bellows WK (2004b) *Collinsiella* (Ulvophyceae, Chlorophyta) and other ulotrichalean taxa with shell-boring sporophytes form a monophyletic clade. *Phycologia* 43:41-49.

O'Kelly CJ, Wysor B & Bellows WK (2004c) Gene sequence diversity and the phylogenetic position of algae assigned to the genera *Phaeophila* and *Ochlochaete* (Ulvophyceae, Chlorophyta). *J. Phycol.* 40:789-799.

Ohta Y & Yonekura N (1987) Explanatory text for quaternary maps of Japan. *In* Japan Association for Quaternary Research, Section 13, Coast-line. University of Tokyo Press, Tokyo, Japan. pp 70-72.

Papenfuss GF (1960) On the genera of *Ulvales* and the status of the order. *J. Linn. Soc. (Bot.)* 56:303-318.

Pascher A (1914) Über Flagellaten und Algen. *Berichte der deutsche botanischen Gesellschaft* 32:136-160.

Pickett-Heaps JD (1969) The evolution of the mitotic apparatus: an attempt at comparative ultrastructural cytology in the dividing plant cells. *Cytobios* 1:257-280.

Reed RH & Russell G (1978) Salinity fluctuations and their influence on "bottle brush" morphogenesis in *Enteromorpha intestinalis* (L.) Link. *British Phycological Journal* 13:149-153.

Schlichting Jr HE (1974) Ejection of microalgae into the air via bursting bubbles. *J. Allergy Clin. Immunol.* 53:185-188.

Sluiman HJ (1985) A cladistic evaluation of the lower and higher green plants (Viridiplantae). *Plant Syst. Evol.* 149:217-232.

Stiller JW & Hall BD (1997) The origin of the red algae: implications for plastid evolution. *Proc. Natl. Acad. Sci. USA*. 94:4520-4525.

Tan IH, Blomster J, Hansen G, Leskinen E, Maggs CA, Mann DG, Sluiman HJ & Stanhope MJ (1999) Molecular phylogenetic evidence for a reversible morphogenetic switch controlling the gross morphology of two common genera of green seaweeds *Ulva* and *Enteromorpha*. *Mol. Biol. Evol.* 16:1011-1018.

Tappan H (1980) *The paleobotany of plant protists*. WH Freeman &Co, New York, USA.

Tatewaki M (1969) Culture studies on the life history of some species of the genus *Monostroma*. Scientific papers of the institute of algological research, Hokkaido University, Japan. 6:1-56.

Thuret G (1854) Note sur la synonymie des *Ulva lactuca* et *latissima* L. suivie de quelques remarques sur la tribu des Ulvacées. *Mém. Soc. Sci. Nat. Cherbourg*. 2:17-32 (in French).

Valiela I, McClelland J, Hauxwell J, Behr PJ, Hersh D & Foreman K (1997) Macroalgal blooms in shallow estuaries: controls and ecophysiological and ecosystem consequences. *Limnol. Oceanogr.* 42:1105-1118.

van den Hoek C, Mann DG & Jahns HM (1995) *Algae: an introduction to phycology.* Cambridge University Press, Cambridge, UK. 623 pp.

Vinogradova KL (1969) K. sistematike poryadka Ulvales (Chlorophyta) s.l. *Bot. Zh. (St. Petersbg.)* 54:1347-1355 (in Russian).

Vinogradova KL (1974) *Ul 'vovye vodorosli (Chlorophyta) morej SSSR*. Nauka, Leningrad, Russia. 166 pp (in Russian).

Watanabe S, Kuroda N & Maiwa F (2001) Phylogenetic status of *Helicodictyon planctonicum* and *Desmochloris halophila* gen. et comb. nov. and the definition of the class Ulvophyceae (Chlorophyta). *Phycologia* 40:421-434.

Wittrock VB (1866) Försök till en monographi öfver algslägtet *Monostroma*. Ph.D. dissertation, Uppsala University, Sweden. 66 pp (in Swedish).

Yamamoto I, Takahashi, M, Tamura, E. & Maruyama, H (1982) Antitumour activity of crude extracts from edible marine algae against L-1210 leukemia. *Bot. Mar.* 25:455-457.

Yoon HS, Hackett J, Ciniglia C, Pinto G & Bhattacharya D (2004) A molecular timeline for the origin of photosynthetic eukaryotes. *Mol. Biol. Evol.* 21:809-818.

Zhang H-J, Mao W-J, Fang F, Li H-Y, Sun H-H, Chen Y & Qi X-H (2008) Chemical characteristics and anticoagulant activities of a sulfated polysaccharide and its fragments from *Monostroma latissimum*. *Carbohydrate polymers* 71:428-434.

In: Algae
Editors: Dagmar Krueger and Helga Meyer
ISBN: 978-1-62081-280-8

Chapter 6

SOIL ALGAE: ECOPHYSIOLOGICAL FEATURES, ENVIRONMENTAL IMPACT AND BIOTECHNOLOGICAL POTENTIAL

Werner Reisser*
University of Leipzig, Institute of Biology, Leipzig, Germany

ABSTRACT

In most biological textbooks the term "alga" is used to designate a group of photosynthetically active organisms that are bound to aquatic habitats and lack the classical differentiation into root, stem and leaves. Only experts are aware of the fact that algae are very successful competitors in terrestrial and aerial habitats as well and that they can survive also under aphotic conditions. Accordingly, there is ample information available on taxonomy, physiology and ecology of aquatic algae and on their application in biotechnology whereas on aeroterrestrial algae comparably few data exist. This is the more astonishing since aeroterrestrial algae have to cope with much more diverse and quickly changing environmental stressors than aquatic algae do that live in a more or less constant environment. Aeroterrestrial algae are specialists for interfaces, they settle e. g. substrates exposed to the air such as soil, tree barks, rocks, fences, roofs, facades and rubber seals of car windows. There environmental conditions such as light, UV, temperature and moisture can change quickly. What is more, those algae contribute a significant part to CO_2-fixation in their ecosystems, the exact amount of which still needs thorough investigation. Thus the study of the ecophysiology of aeroterrestrial algae, of their adaptive features and their potential biotechnological applicability is one of the most promising fields in current phycological research.

In this chapter an overview will be given on what is known on the ecophysiology of soil algae, i. e. on that part of the aeroterrestrial algae that lives in soil, and on current research on their environmental impact as well as on their potential use in biotechnology.

* E-mail: reisser@rz.uni-leipzig.de.

INTRODUCTION

In current discussion on the ecological impact of global climate change the consequences as to soil ecosystems appear to be the least predictable. As a matter of fact soil is the least understood part of the biosphere although the overwhelming part of both human and animal food supply depends on this limited source (Banwart 2011; Foley et al. 2011; Schuur and Abbott 2011). Soil organic matter contains more than three times as much carbon than either the atmosphere or the terrestrial vegetation (Schmidt et al. 2011) but its role in carbon storage is still far from being settled. Recent data show (Schmidt et al. 2011) that stability and turnover rates of soil organic matter depend to a much greater extent than thought before on the activity of soil organisms (Davidson and Janssens 2006; Knorr et al. 2005; Powlson 2005; Schuur and Abbott 2011; Yergeau et al. 2011) Thus a better knowledge on both activity and life cyles of soil organisms is an important prerequisite to predict the future reaction of soil ecosystems.

In general, only limited information is available on the role of various organisms in soil carbon fluxes and hence on the consequences of their reaction to a change of environmental parameters such as temperature, humidity, pH, etc. This information gap holds also for algae living in instead of on soil. Up to now only a limited and rather scattered and incoherent set of data deals with those genuine soil algae. Neither their interaction with other soil organisms nor their influence on plant germination and growth is known. Information on their metabolic activity, their persistence in soil under aphotic conditions and hence their share in general deep soil carbon turnover is limited.

Therefore this review aims at giving an overview on current data on soil algae with preference to the aspects discussed above and to their potential use for biotechnological purposes. As soil algae only those eukaryotic algae will be dealt with that are living in soil, i. e. mainly in the aphotic zone. Crust building algae will be deliberately excluded, also those algae of extreme habitats such as hot or cold deserts since they all have their own expert clientele and are well documented in literature.

NUMBERS, BIOMASS AND TURNOVER

Numbers of soil algae are easily determined by either direct cell counting with a microscope that may be additionally equipped with a UV source for obtaining chlorophyll fluorescence (Tchan 1952) or by plating techniques (Stellmacher and Reisser 1999). While plating techniques give information on the amount of viable cells, observation of chlorophyll fluorescence includes also resting stages and gives a preliminary information on the morphology of observed algae.

A general problem in trying to obtain basic data on cell numbers of soil algae, their biomass and its turnover is the great heterogeneity of the habitat they are living in. Soil per se is rather inhomogeneous consisting of an enormous variety of microhabitats that differ on a small scale and offer quite different chances for algal growth. Thus algal counts within an area of one square meter may vary tremendously depending on season (Grondin and Johanson 1995) but also on day time and grazing pressure (Tchan and Whitehouse 1953), on the covering of the soil surface by e. g. leaves, the presence of roots or fertilizers (Zancan et al.

2006), acid rain (Lukesova and Tajkovsky 1999), pollution by heavy metals (Kalinowska et al. 2004), organic pollutants (Megharaj et al. 2000b), etc. Thus any upscaling of cell numbers to an acre or an hectar and estimation of biomass has to be done with extreme caution.

The same caveats hold for any estimation of physiological activity and turnover of biomass. Only part of soil algae lives in the photic zone or has access to a sufficient water supply.

In addition, turnover of algal biomass may also depend on other factors such as the morphology of algae themselves. Thus coccoid forms are easily washed into deeper layers of soil and into groundwater whereas filamentous specimen rest for a longer time at the surface of soil (Reisser, unpublished results). The impact of specific grazers as well as the effects of bioturbation in general is not known.

Highest numbers of algae are reported for the surface of agricultural soils (3,3 x 10^7 algal cells per g dw of topsoil, Hunt et al. 1979) where number of viable algae decreases to about 10 % at a depth of 1 m (Stina 1969). Shimmel and Darley (1985) count for an agricultural soil 5,85 x 10^8 algal cells per m^2 at maximum. Hahn and Neuhaus (1997) report about 1,6 x 10^5 algal cells per g dw, 16 % of which are diatoms. In the upper layer of forest soils about 1,6 x 10^6 algal cells per g dw were counted (Hunt et al. 1979).

As to algal biomass up to 150 g per m^2 are given for agricultural soils (Österreicher 1990), for a dry meadow 0,3 g per m^2 and 29,0 g per m^2 for a floodplain meadow are reported (Kabirov and Gaisina 2009). Data given for a pine plantation are 18,8 g x m^{-2} for biomass of soil algae with an average monthly production of 33,8 g x m^{-2}, for a mountain forest (South Ural) according data are 2,3 g x m^{-2} for biomass and 8,1 g x m^{-2} for monthly production (Kabirov and Gaisina 2009).

Shimmel and Darley (1985) calculate the biomass production of soil algae as 39 g C x m^{-2} x a^{-1}. According to Ramsey and Ball (1983) the algal biomass in a pasture soil (1 – 5 cm) amounts to about 5 % of the total microbial biomass.

Unfortunately data give no clue to the relative share of autotrophic and heterotrophic growth in biomass production. However, given the fact that part of the algal population almost certainly is in a resting stage, it is obvious that a significant part of biomass is produced by algae living in the aphotic region.

TAXONOMY

The taxonomy of soil algae has been studied thoroughly by optical methods such as light and electron microscopy and by studies of the cell cycle of isolates (Bold 1970; Brendemühl 1950; Ettl and Gärtner 1995; Johansen and Shubert 2001; Lund 1962; Metting, 1981). In general, chlorophyceaen, xanthophyceaen and bacillariophyceaen taxa are dominating, eustigmatophyceaen, euglenophyceaen, dinophycean and rhodophycean genera have also been reported, but at lower numbers. There seems to be some predominance of chlorophyceaen algae in acid soils (Lukesova and Hoffmann 1995, 1996; Starks et al., 1981). In agricultural soils about 16 % of species found belong to Diatoms (Hahn and Neuhaus 1997). Most studies concentrate on special habitats, such as agricultural soils (Hahn and Neuhaus 1997; Stina 1969), forest soils (Grondin and Johansen 1995; Hunt et al. 1979; Komaromy 1976; Wöhler et al. 1999), sandy soils (Komaromy 1979), alpine soils (Gärtner and Ingolic 1993; Pitschmann 1969; Reisigl 1964, 1968; Trenkwalder 1975; Vinatzer 1975), hot and cold desert soils (Cavacini 2001), and disturbed and reclamation soils (Johansen et al. 1982). Other reports concentrate on soils of special geographic regions such as Japan (Nakano 1990; Watanabe 1977a, 1977b), Papua New Guinea (Watanabe 1979), New Zealand (Ramsey and Ball 1983), Hawaii (Carson and Brown 1978), Texas (Deason and Bold 1960), England (Broady 1979), Connecticut (Hilton and Trainor 1963), Florida (Ruble and Davis 1988).

To sum up, studies suggest that there are ubiquitous taxa such as *Chlamydomonas* spp., *Stichococcus* spp. and *Klebsormidium* spp. as well as taxa showing some preference for special habitats. Thus Bednarz and Starzecka (1990) report on a *Pseudococcomyxa simplex* forming a monoculture in a forest ecosystem that was exposed to a cadmium dose of about 2000 t per square kilometer and year.

Another caveat as to the actual knowledge of the taxonomy of soil algae is given by the fact that the overwhelming number of taxonomic affiliations rest exclusively on microscopical techniques. But as has been pointed out by Rindi et al. (2009) there is ample evidence for the morphological convergence of not related taxa, mostly in unicellular and simple filamentous forms. Hence additional methods are urgently required. An interesting attempt is made by Görs et al. (2010) who use ergosterol and by Gustavs et al. (2011) who apply polyols as chemotaxonomic markers for identifying chlorophyceaen algal species.

In a key paper on chlorophyceaen algae from desert soils Lewis and Lewis (2005) use 18sRNA data for a taxonomic affiliation and conclude that the observed desert green algae – although being similar in morphology – are quite different from aquatic specimen and document a separate path of evolution. This sheds an interesting new light on the old discussion whether there exist endemic taxa among soil algae. Common wisdom in most textbooks is that there is a permanent influx of algae transported by air into the soil ecosystem. Most of them are taken away again by wind or water driven erosion but others may penetrate in deeper soil layers and be identified there as “soil algae”. Algae coming in are probably of quite diverse origin, stemming from fresh or salty waters, from other soil habitats or from algal crusts on barks, walls, fences, etc.

Ecophysiology and Influence of Soil Algae on Their Ecosystem

Current knowledge on the ecophysiology of soil algae is rather limited and is mainly derived from experiments on the reaction of algae in disturbed or contaminated ecosystems and from indirect conclusions. Basic data on photosynthetic and respiratory activity are lacking. This reflects again the general problem of handling the premises of a most heterogeneous habitat. Accordingly, most data are available from algae living in extreme habitats such as cold or hot desert soils where a change in environmental parameters is limited and predictable. In those soils algae are the most important primary producers. However, the question remains open for discussion, whether under aphotic conditions soil algae can grow actively or just exist as resting stages with life times dependent on environmental conditions.

It has been shown repeatedly that resting stages of soil algae can survive in a desiccated status up to 80 years (Metting 1981; Trainor 1970, 1985) and after rewetting can recover and form flagella within a couple of minutes. Thus formation of resting stages may be a common survival strategy of soil algae allowing them to endure a wide temperature range as met in permafrost soils (Vishnivetskaya et al. 2003) and in hot deserts with temperatures up to 50 °C (Buzer et al. 1985; Trainor 1983).

Corresponding to temperature water supply changes and might act as a further stressor. A reduction of the amount of available water usually means an increase of salinity. As is known from marine algae there exist various countermeasures to osmotic stress such as the production of sugar alcools and glycerol (Kirst 1989). As a matter of fact in some soil algae the production of polyols could be observed under osmotic stress. However, this is not a common feature, e. g. in *Klebsormidium flaccidum*, one of the most common soil algae, production of polyols could not be observed (Reisser, unpublished results).

Some soil algae are able to produce exopolysaccharides as slime or capsules that form a sort of water storing envelope around cells (Kroen 1984; Kroen et Rayburn 1984; Hokputsa et al. 2003; Paulsen et al. 1992) but their role in survival strategies still needs experimental confirmation.

Besides water the second important environmental parameter influencing the well being of a soil alga is light. It has been shown repeatedly that the photic zone, i. e. the zone in which sufficient light required for autotrophic growth is available, is rather small. Given that the surface of soil is not covered and open to sunlight, as is usually the case in agricultural soils, most algae are below the photosynthetic compensation point at depths of 1 cm (Tchan and Whitehouse 1953; Reisser, unpublished results), at maximum at 4 – 5 cm (Tester and Morris 1987). However, it is an interesting and repeatedly confirmed observation that the absolute number of viable soil algae in the photic zone generally is lower than in the aphotic zone. As has been shown e.g. by Stina (1969) for an agricultural soil the number of viable algal cells in the upper layers was about 1, 0 x 10^3 cells per g soil, in a depth of 20 cm also about 1,0 x 10^3 cells, in a depth of 60 cm about 0,25 x 10^3 and in 100 cm about 0,1 x 10^3 cells. This means that there is a constant input of live algal cells from the upper layers into the deeper layers of soil due to transport by rain or e.g. perturbation by grazers and other animals. The main and still unsolved question is whether those algae living in the aphotic zone are still actively growing or just trying to survive as a resting stage. Active growth would require a switch to organic C-sources, i. e. heterotrophic growth, and a sufficient amount of oxygen for respire-ation unless fermentation is taken into account.

On fermentative activities of soil algae resilient data are not available. Respiration of organic C-sources such as mono- and disaccharides are reported for some soil algal taxa (Bristol-Roach 1926, 1927, 1928; Metting 1981; Parker 1961) but a systematic survey is lacking. Tchan and Whitehouse (1953) observe, that in sandy soils algal growth is inhibited when the water content surpasses a certain threshold. This finding could indicate that the oxygen supply may also be a limiting factor to algal respiration. Bednarz and Starzecka (1990) estimate the algal respiration in acid forest soils to be less than 1 % of total soil respiration but a thorough study on respiration of algae in different type of soils is lacking.

In contrast to the lack of basic studies on soil algal metabolism there are far more studies on the ecology of soil algae in general (Hunt et al. 1979; Kabirov and Gaisina 2009; King and Ward 1977; Starks et al. 1981), their influence on the physical properties of soil (Bond and Harris1964) and on their role as pioneering organisms settling new habitats or as biomonitors for changing environmental parameters.

Specific reaction of *Klebsormidium* spp. to changing environmental conditions is studied by Jahnke et al. (2011) and Karsten and Holzinger (2011). Pipe and Shubert (1984), Shields and Durrell (1964) and Zancan et al. (2006) report on soil algae as indicators for soil fertility. The reaction of the soil algal population to contamination of their habitat by copper is studied by Kalinowska and Pawlik-Skowronska (2010) and by Kalinowska et al. (2004), to cadmium by Bednarz and Starzecka (1990), to Pb and Zn by Cabala et al. (2011), to oil pollution and further organic compounds in general by Ghasemi et al. (2011), Megharaj et al. (2000b) and Safonova et al. (2005). In this context soil algae have been tested repeatedly for biomonitoring tasks, such as for the contamination of soils by herbicides (Eladel et al. 1999; Megharaj et al. 1999, 2000a; Neuhaus and Seefeld 2000; Neuhaus et al. 1997, 2000; Zurek 1981) and for the impact of acid rain (Kabirov and Shilova 1994; Lukesova and Hoffmann 1995, 1996).

Soil algae and their role in settling new habitats have been studied by Carson and Brown (1978), by Rayburn et al. (1982), and by Schwabe and Behre (1972) on volcanic substrates, by Lukesova (2001) and Lukesova and Komarek (1987) and Starks and Shubert (1978) on dumps of mining areas, and by Johansen et al. (1982) on burned areas.

The ability of soil algae to settle new habitats depends not least on the production of extracellular substances such as exopolysaccharides that aggregate soil particles, protect against erosion by wind or water, keep the moisture and thus help in forming new soils or in protecting existing ones (Bailey et al. 1973; Hu et al 2002). Thus they also facilitate succession and the following settlement of lichens and mosses.

There is also growing evidence that at least some soil algal species are able to produce substances that may help them to compete with other soil organisms such as bacteria and fungi. Thus the excretion of exoenzymes such as chitinases and laccases (Otto et al. 2011) has been reported as well as the production of antibiotics (Harder and Oppermann 1953; Safonova and Reisser 2005). Soil algae may be able to influence plant growth by production of phytohormones (Tarakhovskaya et al. 2007) and also growth of bacteria by interfering with their quorum sensing mechanisms (Teplitski et al. 2011).

Last but not least soil algae are an important basis of terrestrial food chains by forming part of the diet of earthworms (Curry and Schmidt 2007; Schmidt et al. 2003), snails (Shachak and Steinberger 1980), tardigrades (Hohberg et al. 2011), mites (Hubert and Lukesova 2001) and various Diplopoda and Oniscidea (Lukesova and Tajkovsky 1999).

Biotechnology with Soil Algae

Since most soil algae can be grouped among the microalgae they participate in the general interest in microalgal products (Pulz and Gross 2004). But there are also various attempts to exploit special ecophysiological features of soil algae for biotechnological use. Besides more general projects such as the search for algal enzymes adapted to extreme (hot or cold, acid or basic) conditions, there is growing evidence, that in soil algae there might exist some interesting enzymes applicable to phytoremediation measures such as laccases (Otto et al. 2010). A lot of studies deal with the use of soil algae for reclamation of soils that are polluted by heavy metals (Kalinowska et al. 2004, 2010; Cabala et al. 2011) or by hydrocarbons due to oil spill (Megharaj et al. 2000b; Safonova et al. 2005). Also reclamation of surface mined areas with the help of soil algae is an important topic (Lukesova 2001; Lukesova and Komarek 1987; Shubert and Starks 1980, 1985; Starks and Shubert 1978, 1982). A further field of interest is the use of algae for phytoremediation of soils contaminated by arsenic compounds (Huq et al. 2007; Pawlik-Skowronska et al. 2004).

As to the reclamation or desert or eroded soils the use of proteoglycan producing algae (Flaibani et al. 1989; Hu et al. 2002; Liang 2006) is discussed.

Last but not least there are various attempts to use soil algae as bioindicators of soil fertility and for biotests (Hammel et al. 1998; Jahnke et al. 2011).

REFERENCES

Bailey, D., Mazurak, A. P. and Rosowski, J. R. (1973). Aggregation of soil particles by algae. *J. Phycol.* 9, 99-101.

Banwart, S. (2011). Save our soils. *Nature* 474, 151-152.

Bednarz, T. and Starzecka, A. (1990). Effect of metallurgic industrial dusts on the development and activity of algae and bacteria in soils of Niepolomice Forest (Southern Poland). *Arch. Hydrobiol.* 87, 65-82.

Bold, H. C. (1970). Some aspects of the taxonomy of soil algae. *Ann. N. Y. Acad. Sci.* 175, 601-616.

Bond, R. D. and Harris, J. R. (1964). The influence of the microflora on the physical properties of soils. I. Effects associated with filamentous algae and fungi. *Australian J. of Soil Research* 2, 111-122.

Brendemühl, I. (1950). Über die Verbreitung der Erddiatomeen. *Arch. Mikrobiol.* 14, 407-449.

Bristol Roach, B. M. (1926). On the relation of certain soil algae to some soluble carbon compounds. *Ann. Bot.* 40, 149-201.

Bristol Roach, B. M. (1927). On the carbon nutrition of some algae isolated from soil. *Ann. Bot.* 41, 509-517.

Bristol Roach, B. M. (1928). On the influence of light and of glucose on the growth of soil algae. *Ann. Bot.* 42, 317-345.

Broady, P. A. (1979). Qualitative and quantitative observations on green and yellow-green algae in some English soils. *Br. Phycol. J.* 14, 151-160.

Buzer, J. S., Dohmeier, R. A. and du Toit, D. R. (1985). The survival of algae in dry soils exposed to high temperature for extended time periods. *Phycologia* 24, 249-251.

Cabala, J., Rahmonov, O., Jablonska, M. and Teper, E. (2011). Soil algal colonization and its ecological role in an environment polluted by past Zn-Pb mining and smelting activity. *Water Air Soil Pollut.* 215, 339-348.

Carson, J. L. and Brown, R. M. (1978). Studies of Hawaiian freshwater and soil algae II. Algal colonization and succession on a dated volcanic substrate. *J. Phycol.* 14, 171-178.

Cavacini, P. (2001). Soil algae from northern Victoria Land (Antarctica). *Polar Biosc.* 14, 45-60.

Curry, J. P. and Schmidt, O. (2007). The feeding ecology of earthworms – A review. *Pedobiologia* 50, 463-477.

Davidson, E. A. and Janssens, I. A. (2006). Temperature sensitivity of soil carbon decomposition and feedback to climate change. *Nature* 440, 165-173.

Deason, T. R. and Bold, H. C. (1960). Phycological studies I. Exploratory studies of Texas soil algae. The University of Texas Publication 6022, 1-72.

Eladel, H. M., Henley, W. J. and Kobbia, I. A. (1999). Effect of thiobencarb on growth and photosynthesis of the soil alga *Protosiphon botryoides* (Chlorophyta). *J. Appl. Phycol.* 10, 547-554.

Ettl, H. and Gärtner, G. (1995). Syllabus der Boden-, Luft- und Flechtenalgen. Stuttgart, Jena, New York: Gustav Fischer.

Flaibani, A., Olsen, Y. and Painter, T. J. (1989). Polysaccharides in desert reclamation: Compositions of exocellular proteoglycan complexes produced by filamentous blue-green and unicellular green edaphic algae. *Carbohydrate Research* 190, 235-248.

Foley, J. A., Ramankutty, N., Braumann, K. A., Cassidy, E. S., Gerber, J. S., Johnston, M., Mueller, N. D., O`Connell, C., Ray, D. K., West, P. C., Balzer, C., Bennett, E. M., Carpenter, S. R., Hill, J., Monfreda, C., Polasky, S., Rockström, J., Sheehan, J., Siebert, S., Tilman, D. and Zaks, D. P. M. (2011). Solutions for a cultivated planet. *Nature* 478, 337-342.

Gärtner, G. and Ingolic, E. (1993). Zur Morphologie und Taxonomie einiger Bodenalgen (Unterfamilie Scotiellocystiodeae, Chlorellaceae) aus der Algensammlung in Innsbruck (ASIB, Austria). *Arch. Protistenk.* 143, 101-112.

Ghasemi, Y., Rasoul-Amini, S. and Fotooh, E. (2011). The biotransformation, biodegradation, and bioremediation of organic compounds by microalgae. *J. Phycol.* 47, 969-980.

Görs, M., Schumann, R., Gustavs, L. and Karsten, U. (2010). The potential of ergosterol as chemotaxonomic marker to differentiate between *Chlorella* species (Chlorophyta). *J. Phycol.* 46, 1296-1300.

Grondin, A. E. and Johansen, J. R. (1995). Seasonal succession in a soil algal community associated with a beech-maple forest in northeastern Ohio, U.S.A. *Nova Hedwigia* 60, 1-12.

Gustavs, L., Görs, M. and Karsten, U. (2011). Polyol patterns in biofilm-forming aeroterrestrial green algae (Trebouxiophyceae, Chlorophyta). *J. Phycol.* 47, 533-537.

Hahn, A. and Neuhaus, W. (1997). Boden-Diatomeen einer landwirtschaftlichen Nutzfläche bei Potsdam, Deutschland. Nova Hedwigia 65, 285-298.

Hammel, W., Streubing, L. and Debus, R. (1998). Assessment of the ecotoxic potential of soil contaminants by using a soil-algae test. *Ecotoxicol. Environ. Saf.* 40, 173-176.

Harder, R. and Oppermann, A. (1953). Über antibiotische Stoffe bei den Grünalgen *Stichococcus bacillaris* und *Protosiphon botryoides*. *Arch. Mikrobiol.* 19, 398-401.

Hilton, R. L. and Trainor, F. R. (1963). Algae from a Connecticut soil. *Plant and Soil* 19, 396-399.

Hohberg, K., Russell, D. J. and Elmer, M. (2011). Mass occurrence of algal-feeding tardigrade, *Apodibius confuses*, in the young soils of a post-mining site. *J. of Zoological Systematics and Evolutionary Research* 49, 62-65.

Hokputsa, S., Hu, C., Paulsen, B. S. and Harding, S. E. (2003). A physico-chemical comparative study on extracellular carbohydrate polymers from five desert algae. *Carbohydrate Polymers* 54, 27-32.

Hu, C., Liu, Y., Song, L. and Zhang, D. (2002). Effect of desert soil algae on the stabilization of fine sands. *J. applied Phycology* 14, 281-292.

Hubert, J. and Lukesova, A. (2001). Feeding of the panphytophagous orbatid mite *Scheloribates laevigatus* (Acari: Oribatida) on cyanobacterial and algal diets in laboratory experiments. *Applied Soil Ecology* 16, 77-83.

Hunt, M. E., Floyd, G. L. and Stout, B. B. (1979). Soil algae in field and forest environments. *Ecology* 60, 362-375.

Huq, S. M. I., Abdullah, M. B. and Joardar, J. C. (2007). Bioremediation of arsenic toxicity by algae in rice culture. *Land Contamination and Reclamation* 15, 327-333.

Jahnke, J., Mahlmann, D. M., Jacobs, P. and Priefer, U. B. (2011). The influence of growth conditions on the cell dry weight per unit biovolume of *Klebsormidium flaccidum* (Charophyta), a typical ubiquitous soil alga. *J. Appl. Phycol.* 23, 655-664.

Johansen, J. R. and Shubert, L. E. (2001). Algae in soils. Nova Hedwigia 123, 297-306.

Johansen, J. R., Javakul, A. and Rushforth, S. R. (1982). Effects of burning on the algal communities of a high desert soil near Wallsburg, Utah. *Journal of Range Management* 35, 589-600.

Kabirov, R. R. and Gaisina, L. A. (2009). Parameters of the productivity of soil algae in terrestrial ecosystems. *Eurasian Soil Science* 42, 1374-1379.

Kabirov, R. R. and Shilova, I. I. (1994). Communities of soil algae in the territory of industrial enterprises. *Russian Journal of Ecology* 25, 409-412.

Kalinowska, R. and Pawlik-Skowronska, B. (2010). Response of two terrestrial green microalgae (Chlorophyta, Trebouxiophyceae) isolated from Cu-rich and unpolluted soils to copper stress. *Environ. Pollut.* 258, 2778-2785.

Kalinowska, R., Pawlik-Sowronska, B. and Skowronski, T. (2004). Cu-adaptation of soil green alga *Stichococcus bacillaris* fo. *minor* isolated from copper polluted grounds. *Acta Physiologiae Plantarum* 26, 205.

Karsten, U. and Holzinger, A. (2011). Light, temperature, and desiccation effects on photosynthetic activity, and drought-induced ultrastructural changes in the green alga *Klebsormidium dissectum* (Streptophyta) from a high alpine soil crust. *Microb. Ecol.* Epub ahead of print, DOI 10.1007/s00248-011-9924-6.

King, J. M. and Ward, C. H. (1977). Distribution of edaphic algae as related to land usage. *Phycologia* 16, 23-30.

Kirst, G.O. (1989). Salinity tolerance of eukaryotic marine algae. *Ann. Rev. Plant Physiol. Plant Mol. Biol.* 40, 21-53.

Knorr, W., Prentice, I. C., House, J. I. and Holland, E. A. (2005). Long-term sensitivity of soil carbon turnover to warming. *Nature*, 433, 298-301.

Komaromy, Z. P. (1976). Soil algal growth types as edaphic adaptation in Hungarian forest and grass steppe ecosystems. *Acta Botanica Academiae Scientarium Hungaricae* 22, 373-379.

Komaromy, Z. P. (1979). Algal flora of Hungarian sandy soils I. Some algological investigations in Kiskunsag national park, Hungary. *Ann. Hist.-nat. Mus. Nat. Hung.* 71, 57-63.

Kroen, W. K. (1984). Growth and polysaccharide production by the green alga *Chlamydomonas mexicana* (Chlorophyceae) on soil. *J. Phycol.* 20, 616-618.

Kroen, W. K. and Rayburn, W. R. (1984). Influence of growth status and nutrients on extracellular polysaccharide synthesis by the soil alga *Chlamydomonas mexicana* (Chlorophyceae). *J. Phycol.* 20, 253-257.

Lewis, L.A. and Lewis, P. O. (2005). Unearthing the molecular phylodiversity of desert soil green algae (Chlorophyta). *Syst. Biol.* 54, 936-947.

Liang, N. (2006). Effect of *Phaeocystis* sp. sprayed over deteriorated soil: a possible method which restores and fertilizes eroding barren land. Diss. Kochi Univ. of Technology, Kochi, Japan.

Lukesova, A. (2001). Soil algae in brown coal and lignite post-mining areas in central Europe (Czech Republic and Germany). *Restoration Ecology* 9, 341-350.

Lukesova, A. and Hoffmann, L. (1995). Soil algae from acid rain impacted forest areas of the Krusne Hory Mts. 2. Effect of pH on growth. *Algological Studies* 78, 39-51.

Lukesova,A. and Hoffmann, L. (1996). Soil algae from acid rain impacted forest areas of the Krusne Hory Mts. 1. Algal communities. *Vegetatio* 125, 123-136.

Lukesova, A. and Komarek, J. (1987). Succession of soil algae on dumps from strip coal-mining in the Most region (Czechoslowakia). *Folia Geobotanica et Phytotaxonomica* 22, 355-366.

Lukesova, A. and Tajkovsky, K. (1999). Interactions between soil algae and saprophageous invertebrates (Diplopoda and Oniscidea). Soil Zoology in Central Europe, Tajkovsky, K., and Pzl, V. (eds.), 187–185 , ISB AS CR.

Lund, J. W. G. (1962). Soil algae, in: Physiology and Biochemistry of Algae, (R. A. Lewin, ed.), 759-770, Academic Press.

Megharaj, M., Boul, H. L. and Thiele, J. H. (1999). Effects of DDT and its metabolites on soil algae and enzymatic activity. *Biology and Fertility of Soils* 29, 130-134.

Megharaj, M., Kantachote, D., Singleton, I. and Naidu, R. (2000a). Effect of long-term contamination of DDT on soil microflora with special reference to soil algae and algal transformation of DDT. *Environmental Pollution* 109, 35-42.

Megharaj, M., Singleton, I., McLure, N. C. and Naidu, R. (2000b). Influence of petroleum hydrocarbon contamination on microalgae and microbial activities in a long-term contaminated soil. *Archives of environmental contamination and toxicology* 38, 439-445.

Metting, B. (1981). Systematics and ecology of soil algae. *Bot. Rev.* 47, 195-312.

Nakano, T. (1990). Some soil algae from the Senkakuku islands, Ryukyus, Japan. *The Journal of Japanese Botany*, 65, 171-180.

Neuhaus, W. and Seefeld, F. (2000). Ökotoxische Auswirkungen der Anwendung von Stomp SC (Pendimethalin) und Fenikan (Diflufenican und Isoproturon) auf die Abundanz terrestrischer Algen in einem lehmigen Sandboden. *Gesunde Pflanzen* 52, 16-25.

Neuhaus, W., Seefeld, F. and Hahn, A. (1997). Auswirkungen von Igran 500 flüssig auf die Abundanz von Bodenalgen unter Labor- und Freilandbedingungen. *Nachrichtenbl. Deut. Pflanzenschutzd.* 49, 260-267.

Neuhaus, W., Malkomes, H. P., Reese-Stähler, G. and Pallutt, B. (2000). Einfluss von Butisan S (Metazachlor) auf Dehydrogenaseaktivität und Kurzzeitatmung sowie Algen im Boden in Beziehung zu den Herbizidrückständen unter Winterraps. *Nachrichtenbl. Deut. Pflanzenschutzd.* 52, 265-273.

Österreicher, W. (1990). Ökologische Bedeutung der Algen im Boden. *Nachrichtenbl. Deut. Pflanzenschutzd.* 42, 122-126.

Otto, B., Schlosser, D. and Reisser, W. (2010). First description of a laccase-like enzyme in soil algae. *Arch. Microbiol.* 192, 759-768.

Parker, B. C. (1961). Facultative heterotrophy in certain soil algae from the ecological viewpoint. *Ecology* 42, 381-386.

Paulsen, B. S., Vieira, A. A. H. and Klaveness, D. (1992). Structure of extracellular polysaccharides produced by a soil *Cryptomonas* (Cryptophyceae). *J. Phycol.* 28, 61-63.

Pawlik-Skowronska, B., Pirszel, J., Kalinowska, B. and Skowronski, T. (2004). Arsenic toxicity and detoxification in the green alga *Stichococcus bacillaris*. *Acta Physiologiae Plantarum* 26, 205.

Pipe, A. E. and Shubert, L.E. (1984). The use of algae as indicators of soil fertility. In: Algae as ecological indicators , Shubert, L. E. (ed.), Academic Press, London, 213-223.

Pitschmann, H. (1969). Zwei neue Heterokonten (*Pleurochloris polyphem* und *Heterococcus clavatus*) aus Böden der alpinen Stufe des Mt. Kenya. *Österr. Bot. Z.* 116, 486-491.

Powlson, D. (2005). Will soil amplify climatic change ? *Nature* 433, 204-205.

Pulz, O. and Gross, W. (2004). Valuable products from biotechnology of microalgae. *Appl. Microbiol. Biotechnol.* 65, 635-648.

Ramsey, A. J. and Ball, K. T. (1983). Estimation of algae in New Zealand pasture soil and litter by culturing and by chlorophyll a extraction. *New Zealand Journal of Science* 26, 493-503.

Rayburn, W.R., Mack, R. N. and Metting, B. (1982). Conspicuous algal colonization of the ash from Mount St. Helens. *J. Phycol.* 18, 537-543.

Reisigl, H. (1964). Zur Systematik und Ökologie alpiner Bodenalgen. *Österr. Bot. Z.* 4, 402-497.

Reisigl, H. (1968). Bodenalgen-Studien II. *Österr. Bot. Z.* 116, 492-506.

Rindi, F., Allali, H. A., Lam, D. W. and Lopez-Bautista, J. M. (2009). An overview of the biodiversity and biogeography of terrestrial green algae. In: Biodiversity Hotspots , V. Rescigno et al. (eds.), 2 – 25.

Ruble, R. W. and Davis, J. S. (1988). Soil algae from fallow potato fields in South Florida (U.S.A.) marl. *Nova Hedwigia* 47, 403-414.

Safonova, E. and Reisser, W. (2005). Growth promoting and inhibiting effects of extracellular substances of soil microalgae and cyanobacteria on *Escherichia coli* and *Micrococcus luteus*. *Phycol. Res.* 53, 189-193.

Safonova, E., Kvitko, K., Kuschk, P., Möder, M. and Reisser, W. (2005). Biodegradation of phenantrene by the green alga *Scenedesmus obliquus* ES-55. *Eng. Life Sci.* 5, 234-239.

Schuur, E. A. G. and Abbott, B. (2011). High risk of permafrost thaw. *Nature* 480, 32-33.

Schmidt, M. W. I., Torn, M. S., Abiven, S., Dittmar, T., Guggenberger, G., Janssens, I. A., Kleber, M., Kögel-Knabner, I., Lehmann, J., Manning, D. A. C., Nannipieri, P., Rasse, D. P., Weiner, S. and Trumbore, S. E. (2011). Persistence of soil organic matter as an ecosystem property. *Nature* 478, 49-56.

Schmidt, O., Dyckmans, J. and Black, K. (2003). Bodenalgen als C-Quelle für Bodentiere. *Mitt. Dtsch. Bodenk. Ges.* **102,** 397–398.

Schwabe; G. H. and Behre, K. (1972). Algae on Surtsey in 1969-1970. The Surtsey Progress Report IV, 85-89.

Shachak, M. and Steinberger, Y. (1980). An algae-desert snail food chain: Energy flow and soil turnover. *Oecologia* 146, 402-411.

Shields, L. M. and Durrell, L. W. (1964). Algae in relation to soil fertility. *Bot. Rev*. 30, 92-128.

Shimmel, S. M. and Darley, W. M. (1985). Productivity and density of soil algae in an agricultural system. *Ecology* 66, 1439-1447.

Shubert, L. E. and Starks, T. L. (1980). Soil-algal relationships from surface mined soils. *Br. Phycol. J.* 15, 417-428.

Shubert, L.E. and Starks, T. L. (1985). Diagnostic aspects of algal ecology in disturbed lands. In: Soil reclamation processes, microbiological analyses and applications, R. L. Tate, D. A. Klein (eds.), Marcel Dekker, pp. 83-106.

Starks, T. L. and Shubert, L. E. (1978). Algal colonization on a reclaimed surface-mined area in western North Dakota. In: M. K. Wali (ed.) *Ecology and Coal Resource development* 2, 652-660.

Starks, T. L. and Shubert, L. E. (1982). Colonization and succession of algae and soil-algal interactions associated with disturbed areas. *J. Phycol*. 18, 99-107.

Starks, T. L., Shubert, E. S. and Trainor, F. R. (1981). Ecology of soil algae: a review. *Phycologia* 20, 65-80.

Stellmacher, G. and Reisser, W. (1999). The quantitative isolation of algae from soil. A suggestion for standardization. *Pedobiologia* 43, 1-3.

Stina, E. A. (1969). Über die Verbreitung und öologische Bedeutung der Algen in Ackerböden. *Pedobiologia* 9, 226-242.

Tarakhovskaya, E. R., Maslov, Y. I. and Shishova, M. F. (2007). Phytohormones in algae. *Russian Journal of Plant Physiology* 54, 163-170.

Tchan, Y. T. (1952). Study of soil algae I. Fluorescence microscopy for the study of soil algae. *Proc. Linn. Soc. NSW* 77, 265-269.

Tchan, Y. T. and Whitehouse, J. A. (1953). Study of soil algae II. The variation of the algal population in sandy soils. *Proc. Linn. Soc. NSW* 78, 160-170.

Teplitski, M. and Rajamani, S. (2011). Signal and nutrient exchange in the interactions between soil algae and bacteria. *Soil Biology* 23, 413-426.

Tester, M. and Morris, C. (1987). The penetration of light through soil. *Plant, Cell and Environment* 10, 281-286.

Trainor, F. R. (1970). Survival of algae in a desiccated soil. *Phycologia* 9, 111-113.

Trainor, F. R. (1983). Survival of algae in soil after high temperature treatment. *Phycologia* 22, 201-202.

Trainor, F. R. (1985). Survival of algae in a desiccated soil: a 25 year study. *Phycologia* 24, 79-82.

Trenkwalder, H. (1975). Neue Bodenalgen aus Föhrenwäldern im Raum von Brixen (Südtirol, Italien). *Ber. nat.-med. Ver. Innsbruck* 62, 7-19.

Vinatzer, V. (1975). Neue Bodenalgen aus den Dolomiten. *Plant Syst. Evol.* 123, 213-235.

Visnivetskaya, T. A., Vorobyova, E. A. and Gilichinsky, D. A. (2002). Viable green algae and cyanobacteria within terrestrial permafrost. *Proceeding of the Second Europen Workshop on Exo/Astrobiology*, 295-298.

Watanabe, S. (1977a). The genus *Chlorella* (Chlorococcales) from Japanese soils. *The Journal of Japanese Botany* 52, 129-136.

Watanabe, S. (1977b). The genus *Chlorella* (Chlorococcales) from Japanese soils (2). *Journ. Jap. Bot.* 52, 129-136.

Watanabe, S. (1979). Green soil algae from Papua New Guinea. In: Studies on cryptogams of Papua New Guinea, S. Kurukawa (ed.), 87-102, Academic Scientific Book Inc., Tokyo.

Wöhler, I. O., Mensching, A. and Reisser, W. (1999). Algae in forest soils: Taxonomic affiliation and distribution. In: N. Rastin and J. Bauhus (eds.): Going underground – ecological studies in forest soils. Research Signpost, Trivandrum, pp. 187-200.

Yergeau, E., Bokhorst, S., Kang, S., Zhou, J., Greer, C. W., Aerts, R. and Kowalchuk, G. A. (2011). Shifts in soil microorganisms in response to warming are consistent across a range of Antarctic environments. *The ISME Journal advance online publication,* doi:10.1038/ismej.2011.124

Zancan, S., Trevisan, R. and Paoletti, M. G. (2006). Soil algae composition under different agro-ecosystems in north-eastern Italy. *Agriculture, Ecosystems and Environment* 112, 1-12.

Zurek, L. (1981). The influence of the herbicides Lenacil and Pyrazon on the soil algae. *Ekologia Polska* 29, 327-342.

In: Algae
Editors: Dagmar Krueger and Helga Meyer
ISBN: 978-1-62081-280-8

Chapter 7

PHENOLIC PROFILE OF CRUDE EXTRACTS DERIVED FROM A RED ALGA, *CORALLINAELONGATA* ELLIS AND SOLANDER

Milagros Rico*[*]*, Aroa López,
Miguel Suárez de Tangil, and Argimiro Rivero
Departamento de Química, Universidad de Las Palmas de Gran Canaria,
Unidad asociada al CSIC, Campus de Tafira,
Canary Islands, Spain.

ABSTRACT

In the present chapter, we have evaluated the total phenolic contents and antioxidant activities of crude extracts from a red alga, *Corallinaelongata*. The phenolic compounds in the extracts have been identified and quantified by reverse-phase high-performance liquid chromatography. Phloroglucinol was by far the predominant polyphenol found in the extracts, followed by catechin and epicatechin.

Extracting solvents such as water, methanol and water/methanol (1/1) significantly affected the total phenolic amounts determined using the Folin-Ciocalteu method. The total phenolic contents ranged from 443.0 to 487.4 mg gallic acid equivalents per 100 g of dry plant material. The antioxidant efficiency of the extracts was evaluated by measuring the ability to scavenge the stable free radical 1,1-diphenyl-2-picrylhydrazyl, finding that it ranged from 37.0 % to 66.4% (inhibition percentage). The high correlation between the yield of extraction, the total phenolic content and the radical scavenging activity was found, allowing us to conclude that phenolic constituents are mainly responsible for the observed antioxidant activity in the extracts. Because of the wide range of biological activities of phloroglucinol and the other polyphenols detected in the extracts, *Corallinaelongate* appears to have many potential industrial uses.

---

[*] Corresponding author: Milagros Rico. Phone: +34 928454418; Fax: +34 9282922, Email: mrico@dqui.ulpgc.es.

INTRODUCTION

Terrestrial and marine plants contain a diverse group of phenolic compounds with the structural requirements of free radical scavengers due to their high reactivity as hydrogen or electron donors and to the ability of polyphenol-derived radicals to stabilize and delocalize unpaired electron (Pietta, 2000, Quideau, Deffieux, Douat-Casassus, Pouysegu, 2011). These compounds have been reported to prevent diseases caused as a result of oxidative stress including cancers, cardiovascular malfunction, cataracts, rheumatism and many other auto-immune diseases besides ageing (Steinmetz and Potter, 1996; Dillar and German, 2000). Therefore, crude extracts of plant materials are increasingly of interest in food and pharmaceutical industries (Southon, 2000).

Seaweeds are considered to be a rich source of phenolic compounds with a number of important biological activities such as phlorotannins, which are polymers of phloroglucinol and constitute an extremely heterogeneous group of molecules (Target and Arnold, 1998; Shibata *et al.*, 2002; Nagayama*et al.*, 2003; Anh*et al.*, 2004; Kang *et al.*, 2004; Yu, Hu, Wu and Liu, 2009). The presence of phloroglucinol is associated to algae containing phlorotannins (Koch and Gregson, 1984; Targett, Boettcher, Targett and Vrolijk, 1995; Shibata *et al.*, 2004). Phloroglucinol was found to inhibit cell damage induced by radiation through scavenging of reactive oxygen species (Kang *et al.*, 2006). Moreover, phloroglucinol showed anti-fouling activity (Lau and Qian, 1997) and therapeutic potential for prevention of chronic inflammation (Kim and Kim, 2010).

Several studies have focused on *Corallinaelongata* algae: with articles on carotenoid composition and xanthophyll regulation in*Corallinaelongata* already published (Esteban *et al.*, 2009); proteins such as R-Phycoerythrin (a protein acting as an accessory photosynthetic pigment in red algae and highly important in many biotechnological applications) also having been isolated from *Corallinaelongata* (Rossano*et al.*, 2003) and with reported metal concentration and structural changes in *Corallinaelongata* from hydrothermal vents (Couto, Neto and Rodríguez, 2010). A study has also been made to determine the quality (as food) of *Corallinaelongata* and some 11 other macrophytes by recording ingestion, absorption and total growth rates of *Paracentrotuslividus* fed unlimited rations of the alga over a 6-month period (Frantzisl and Gremaree, 1992). Despite of low absorption rates, the echinoids fed *C. elongata* showed high growth rates (an index of food quality). However, no information of the phenolic profile of *C. elongata* has been reported to date.

The present chapter focuses on identifying the phenolic profile of *C. elongata* and the relationships between antioxidant activities and the phenolic profile of several extracts derived from *C. elongate*. Differences in the phenolic profiles of the extracts were determined by simultaneous identification and quantification of 14 polyphenols (gallic acid, catechin, epicatechin, rutin, p-coumaric acid, myricetin, quercetin and protocatechuic, vanillic, caffeic, ferulic, chlorogenic, syringic and gentisic acids) using reverse-phase high-performance liquid chromatography with a diode array detector (Lopez, Rico, Rivero and Suarez de Tangil, 2011). Given the importance of phloroglucinol and its widely demonstrated activity and presence in algae, we also determined the amounts of phloroglucinol in the extracts.

MATERIALS AND METHODS

Chemicals

Methanol (of HPLC grade) was obtained from Panreac (Barcelona, Spain). Water was purified on a Milli-Q system from Millipore (Bedford, MA, USA). Formic acid provided by Merck (Darmstadt, Germany) was of analytical quality. Folin-Cicolteau`s phenol reagent, sodium carbonate and 1,1-Diphenyl-2-picrylhydrazyl (DPPH) were from Sigma-Aldrich Chemie (Steinheim, Germany).

Polyphenol standards were purchased as follows: gallic acid, protocatechuic acid, chlorogenic acid, (-) epicatechin, quercetin, myricetin, ferulic acid, *p*-coumaric acid, vanillic acid, syringic acid, (+) catechin, phloroglucinol, from Sigma-Aldrich Chemie (Steinheim, Germany); rutin and gentisic and caffeic acids from Merck (Darmstadt, Germany).

Plant Material

The red alga *C. elongata* was freshly collected from the Canary Islands, Spain, at 0-0.1 m depth between February and March 2009. Soon after collection, the algae were rinsed carefully in fresh seawater and then frozen. The frozen samples were lyophilized, pulverized using a blender (Moulinex, 600 W, France) and were kept in darkness at -20ºC under nitrogen.

Preparation of the Seaweed Extracts

Dry powders (5.0 g) were extracted by stirring at room temperature for 3 h using 1.5 ml acetic acid and 48.5 ml of one of the following solvents: absolute methanol, water, water/methanol (1/1).Each extract was filtered for removal of the alga particles. After centrifugation at 7600 g for 10 min, the supernatant was collected and filtered through 0.45 µm filter paper and stored at 4 ºC.

Yield Determinations

Extraction solutions (5 ml) were evaporated, the residues weighed and the amount of extractable substances of each extract expressed as a percentage by weight of the dry alga powder.

Determination of Total Phenolic Contents

The total phenolic content of each extract was estimated using the Folin-Ciocalteu reagent where values were expressed as mg of gallic acid equivalent to 100 gram dry alga powder (Julkunen-Tiitto, 1985). Sample solutions (100 µl) were mixed with 8.4 mL of H_2O

and 0.5 ml of Folin-Ciocalteu reagent. One ml of 20 % Na_2CO_3 was added. After 1 h incubation in darkness at room temperature, the absorbance was measured at 765 nm using a SHIMADZU 1700 UV-Vis spectrophotometer. The estimation of phenolic compounds was carried out in triplicate, and the results averaged. A calibration curve of gallic acid (ranging from 50 to 900 μg/ml) was prepared and the results were determined using the regression equation of the calibration curve ($y = 0.00029x - 0.00025$; $r^2 = 0.9992$).

Free Radical Scavenging Activity on DPPH

The antioxidant activity was measured in terms of radical scavenging capacity, using the DPPH assay (Chu, Chang and Hsu, 2000). One ml of a methanol solution of 0.1 mM DPPH (1,1-diphenyl-2-pikryl-hydrazin) was mixed rapidly into the sample solution (100 μl). The decline in absorbance was recorded at 515 nm against a methanol blank over a period of 16 min. The scavenging activity (%) (RSA) on DPPH radicals was calculated using the equation: RSA = 100 (1-Abs in the presence of sample/Abs in the absence of sample).

Determination of the Phenolic Profile Using HPLC

The dried algae powder (100 mg) was mixed and homogenized using a vortex for 30 s with 0.6 ml of the following: methanol, water and water/methanol (1/1). The mixture was stirred in a rotary shaker for 60 min at room temperature in darkness. After centrifugation at 7000 g for 20 min at 4ºC, the supernatant was colleted and filtered through a 45 μm nylon syringe filter prior to injection. Chromatographic analysis was performed using a Varian ProStar 210 system, equipped with a vacuum de-gasser, binary pump, thermostat column compartment, diode array detector (DAD), a reverse phase Pursuit XRs C18 (250 mm×4.6 mm, 5 μm) column and a Pursuit XRs C18 (10 mm×4.6 mm, 5 μm) guard column (Varian, Barcelona). The separation of gallic acid, (+) catechin, chlorogenic acid, protocatechuic, gentisic and vanillic acids, (-) epicatechin, ferulic acid, myricetin, caffeic acid, rutin, syringic acid, *p*-coumaric acid and quercetin was carried out using a method previously reported by us (Lopez *et al.*, 2011). A gradient system was used involving two mobile phases. Eluent A was water with 0.1% formic acid and eluent B methanol. The flow rate was 1.0 ml/min, and the injection volume was 60 μl of crude extracts. The system operated at 27ºC. The elution conditions applied were: 0-5 min, 20%B isocratic; 5-30 min, with a linear gradient from 20% to 60% B; 30-35 min, 60%B isocratic; 35-40, with a linear gradient from 60% to 20% B and finally, washing and reconditioning of the column. Simultaneous monitoring was set at 270 nm (gallic acid, protocatechuic acid, (+) catechin, vanillic acid, (-) epicatechin and syringic acid), 324 nm (chlorogenic acid, gentisic acid, caffeic acid, *p*-coumaric acid and ferulic acid) and 373 nm (rutin, myricetin, and quercetin) for quantification. The retention times (RT) were as follows: gallic acid (RT: 5.3 min), protocatechuic (RT: 10.0 min), catechin (RT: 12.7 min), chlorogenic acid (RT: 14.9 min), gentisic acid (RT: 17.1 min), vanillic acid (RT: 17.7 min), epicatechin (RT: 17.9 min), caffeic acid (17.9 min), syringic acid (RT: 18.9 min), coumaric acid (RT: 23.4 min), rutin (RT: 28.1 min), ferulic acid (RT: 24.3 min), myricetin (RT: 30.6 min) and quercetin (RT: 34.6 min). To quantify phloroglucinol in the extracts, five different solutions containing phloroglucinol with concentrations in the range of 1.0-200 μg/ml were injected in triplicate establishing the calibration curve by plotting the peak areas versus the concentration of each solution (y = 7149.1x + 28800). The gradient elution profile was as follow: 0-5 min, 20% B isocratic; 5-10 min, with a linear gradient from 20% to 60% B; 10-17 min, linear gradient from 60% to 20% B and finally, washing and re-conditioning of the column. Monitoring was set at 266 nm and the retention time was 7.05 min. The limit of detection (LOD) was found to stand at 0.0014 μg/ml, with the limit of quantification (LOQ) at 0.0046 μg/ml and the recovery 98.42 ± 6.92 (%). The injections were performed in duplicate.

RESULTS

Yield

Solvents such as methanol, ethanol, acetone, chloroform and water have been commonly used for the extraction of phenolics from brown and red algae (Duan, Zhang, Li and Wang, 2006; Yuan and Walsh, 2006). Yields of several extracts derived from *C. elongata* were evaluated as presented in Table 1. The highest yield was obtained using water as solvent (7.95 %) whereas the methanol and aqueous methanol gave similar yields. The results fully align previous studies where extraction yield was found to be strongly dependent on the solvent polarity and extracts prepared with polar solvents were found to give the highest percentages of extractable substances (Hayouni, Abedrabba, Bouix and Hamdi, 2007). As compared to our results, the results of Ganesan, Chandini and Bhaskar (2008) offer lower yields of crude methanol extracts of three red seaweeds (5.01 %, 3.98 % and 2.85 %).

Table 1. The effects of different solvents on the yield, the total phenolic content (TPC) and the radical scavenging activities (RSA) in *Corallinaelongata* algae extracts

Solvent	Yield[a]	TPC[a,b]	RSA[a,c]
Water	7.95 ± 0.21	487.4 ± 0.87	66.4 ± 0.73
Water methanol	6,52 ± 0,22	464.2 ± 1.52	57.5 ± 0.82
Methanol	6,29 ± 0,25	443.0 ± 4.01	37.0 ± 0.47

[a] Means ± standard deviation of three measurements.
[b] Values are expressed as mg gallic acid equivalents per 100 gram of dry alga power.
[c] Values are expressed as % inhibition.

Total Phenolic Contents

From the results shown in Table 1, it is evident that the recovery of phenolic compounds was dependent on the solvent used and its polarity. The order of TPC from high to low was: water (487.4) > water/methanol (464.2) > methanol (443.0). Similar findings were also reported by Kuda, Tsunekawa, Goto and Araki(2005). Our results in Table 1 show high correlation between the total extractable content and the total phenolic content, in line with previously published results (Chew, Lim, Omar, Khoo, 2008).

Earlier research investigations with respect to the crude methanol extracts of three different brown seaweeds gave a total phenolic content of under 290 mg gallic acid equivalents per 100 g seaweed on a dry weight basis (Chandini, Ganesan and Bhaskar, 2008), values inferior to those found here.

The DPPH Radical-Scavenging Activity

The order of the radical-scavenging activity in the extracts from high to low was: water > water/methanol > methanol (Table 1). The higher extraction yield corresponds to the higher antioxidant activity. The water extract showed the highest amount of total phenolic compounds (487.4) and the strongest DPPH radical scavenging activity (66.4%), whereas the methanol extract gave the lowest activity (37.0). A high correlation was found between TPC and RSA. The present findings corroborate well with earlier studies on algae extracts where it was concluded that high levels of total phenolic contents gave high RSA values indicating that phenolic compounds were the major constituents contributing to the DPPH scavenging activities of the extracts (Duan*et al.*, 2006; Chew *et al.*, 2008; Yangthong, Hutadilok-Towatana, Phromkunthong, 2009).

Table 2. Polyphenol amounts in extracts derived from *Corallinaelongata*algae determined by HPLC

Compounds	Solvent		
	Methanol[a] [b]µg / 100 g	Water/methanol[a] [b]µg / 100 g	Water[a] [b]µg / 100 g
Gallic acid	643.5 ± 16.94	426.4 ± 36.96	1700 ± 119.1
Protocatechuic	573.5 ± 11.68	752.4 ± 70.37	704.8 ± 21.6
Catechin	3518 ± 5.208	3699 ± 246.2	4444 ± 125.6
Vanillic acid	630.3 ± 37.59	558.2 ± 29.22	1121 ± 12.19
Epicatechin	1794 ± 99.21	1474 ± 82.69	7349 ± 697.8
Syringic acid	809.9 ± 70.83	769.3 ± 17.02	987.2 ± 98.82
Chlorogenic acid	1102 ± 43.78	1068 ± 35.91	1058 ± 17.19
Gentisic acid	1069 ± 80.32	1110 ± 54.91	1003 ± 21.08
Caffeic acid	374.3 ± 40.77	387.5 ± 35.40	409.0 ± 37.38
Coumaric acid	435.3 ± 42.34	412.9 ± 11.01	471.6 ± 28.07
Ferulic acid	638.5 ± 7.663	623.1 ± 16.64	696.8 ± 4.70
Rutin	501.4 ± 46.66	440.7 ± 26.29	566.3 ± 33.83
Quercetin	nd[c]	nd[c]	nd[c]
Phloroglucinol	nd[c]	17,64 ± 110.0	24,82 ± 70.04
Total amount of phenolics (mg per 100g of dry alga power)			
Sum	13.18	30.50	46.46
Sum - phloroglucinol	13.18	12.89	21.64

[a] Means ± standard deviation of three measurements.

[b] µg / 100 g indicates per 100 gram of dry algae powder.

[c] nd indicates not detected.

The Determination of Phenolic Profile by HPLC

Each extract showed the presence of all the standards with the exceptions of quercetin, which was not to be detected (Table 2). In this analysis, phloroglucinol was by far the predominant polyphenol in the water and water/methanol extracts (24.8 and 17.6 mg per 100 gram dry alga respectively), although it was not to be detected in the methanol extract. Significant amounts of catechin and epicatechin were detected in all the extracts. As compared to our results, Onofrejová*et al.* (2010) reported lower amounts of protocatechuic, *p*-coumaric, vanillic, caffeic and chlorogenic acids extracted from *in vitro* culture of two freshwater algae and from food products of marine macroalgae. High correlation was found between the total phenolic content calculated as the sums of the quantified by HPLC polyphenols (Table 2) and the radical scavenging activity. It must be noted that this high correlation depends strongly on the amount of phloroglucinol. As a general rule, the water gave higher contents of each polyphenol than the other solvents, thereby explaining why the water extract showed higher RSA.

CONCLUSION

Several reports have convincingly shown a close relationship between antioxidant activity and total phenolic content in algae extracts and in other plant materials, concluding that phenolic compounds are responsible for antioxidant activity (Duan*et al.*, 2006; Chew *et al.*, 2008; Alothman, Bhat and Karim, 2009).

Chemical assays based on the removal of stable DPPH free radicals have been shown to indicate the presence of reductive compounds in terms of hydrogen donation capacity under test conditions. The radical scavenging activity of the extracts increased when the total phenolic content increased. These findings indicated that phenolic compounds are the main microconstituents contributing to the antioxidant activity of the algal extracts.

The results of the present work indicate that extracts of *Corallina elongate* algae can be used as a source of phloroglucinol. Phloroglucinol has been reported to be an inhibitory of *Hydroideselegans* larval settlement and the growth of certain marine bacteria, the two mechanisms that have been suggested to explain the antifouling effects of natural compounds (Lau and Qian, 1997). Therefore, aqueous extracts of *Corallinaelongata* can be used as an easy and accessible antifouling agent in coastal areas. In addition, phloroglucinol was found to have many biotechnological applications in food science and therapy (Kang *et al.*, 2004; Kang *et al.*, 2006; Kim and Kim, 2010).

ACKNOWLEDGMENTS

Financial support was given by the Caja Insular de Ahorros de Canarias and is hereby acknowledged. The authors would like to thank Miss Begoña Rocha Sosa for providing algae.

References

Ahn, MJ; Yoon, KD; Min, SY; Lee, JS; Kim, JH; Kim, TG; Kim, SH; Kim, NG; Huh, H and Kim, JW. Inhibition of HIV-1 reverse transcriptase and protease and protease by phlorotannins from the brown algae Ecklonia cava. *Biological and Pharmaceutical Bulletin*, 2004, 27, 544–547.

Alothman, M; Bhat, R and Karim, AA. Antioxidant capacity and phenolic content of selected tropical fruits from Malaysia, extracted with different solvents. *Food Chemistry*, 2009, 115, 785-788.

Chandini, SK; Ganesan, P. and Bhaskar, N. In vitro antioxidant activities of three selected brown seaweeds of India. *Food Chemistry*, 2008, 107, 707-713.

Chew, YL; Lim, YY; Omar, M. andKhoo, KS. Antioxidant activity of three edible seaweeds from two areas in South East Asia. *Food Science and Technology*, 2008, 41, 1067-1072.

Chu, YH; Chang, CL and Hsu, HF. Flavonoid content of several vegetables and their antioxidant activity. *Journal of the Science of Food and Agriculture*, 2000, 80, 561-566.

Couto, RO; Neto, AI and Rodriguez, AS. Metal concentration and structural changes in *Corallina elongate* from hydrothermal vents. *Marine Pollution Bulletin*, 2010, 60, 509-514.

Dillard, CJ and German, JB. Phytochemicals: nutraceuticals and human health. *Journal of the Science of Food and Agriculture*, 2000, 80, 1744-1756.

Duan X-J; Zhang, W-W; Li X-M and Wang, B-G. Evaluation of antioxidant property of extract and fractions obtained from a red alga, *Polysiphoniaurceolata*. *Food Chemistry*, 2006, 95, 37-43.

Esteban, R; Martinez, B; Fernandez-Marin, B; Becerril, JM and Garcia-Plazaola, JI. Carotenoid composition in Rhodophyta: insights into xanthophyll regulation in *Corallina elongate*. *European Journal of Phycology*, 2009, 44, 221–230.

Frantzisl, A andGremaree, A. Ingestion, absorption, and growth rates of *Paracentrotuslividus*(Echinodermata: Echinoidea) fed different macrophytes. *Marine Ecology Progress Series*, 1992, 95, 169-183.

Ganesan, P; Chandini, SK and Bhaskar, N. Antioxidant properties of metanol extract and its solvent fractions obtained from selected Indian red seaweeds. *Bioresource Technology*, 2008, 99, 2717-2723.

Hayouni, EA; Abedrabba, M; Bouix, M and Hamdi, M. The effects of solvents and extraction method on the phenolic contents and biological activities *in vitro* of Tunisian *Quercuscoccifera*L. and *Juniperusphoenicea* L. fruit extracts. *Food Chemistry*, 2007, 105, 267-273.

Julkunen-Tiitto, R. Phenolic constituents on the leaves of northern willows: methods for the analysis of certain phenolics. *Journal of Agricultural and Food Chemistry*, 1985, 330, 213-217.

Kang, KA; Lee, KH; Chae, S; Zhang, R; Jung, MS; Ham, YM; Baik, JS; Lee, NH. and Hyun, JW. Cytoprotective effect of phloroglucinol on oxidative stress induced cell damage via catalase activation. *Journal of Cellular Biochemistry*, 2006, 97, 609-620.

Kang, HS; Chung, HY; Kim, JY; Son, BW; Jung, HA. and Choi, JS. Inhibitory phlorotannins from the edible brown alga Eckloniastolonifera on total reactive oxygen species (ROS) generation. *Archives Pharmacal Research*, 2004, 27, 194–198.

Kim, M-M. and Kim, S-K. Effect of phloroglucinol on oxidative stress and inflammation. *Food and Chemical Toxicology*, 2010, 48, 2925-2933.

Koch, M and Gregson, RP. Brominated phlorethols and nonhalogenatedphlorotannins from the brown alga *Cystophoracongesta*. *Phytochemistry*, 1984, 23, 2633-2637.

Kuda, T; Tsunekawa, M; Goto, H. and Araki, Y. Antioxidant properties of four edible algae harvested in the Noto Peninsula, Japan. *Journal of Food Composition and Analysis*, 2005, 18, 625-633.

Lau, SCK. and Qian, P-Y.Phlorotannins and related compounds as larval settlement inhibitors of the tube-building polychaeteHydroides*elegans*. *Marine Ecology Progress Series*, 1997, 159, 219-227.

Lopez, A; Rico, M; Rivero, A. and Suarez-Tangil, M. The effects of solvents on the phenolic contents and antioxidant activity of *Stypocaulonscoparium* algae extracts. *Food Chemistry*, 2011, 125, 1104–1109.

Nagayama, K; Shibata, T; Fujimoto, K; Honjo, T. and Nakamura, T. Algicidal effect of phlorotannins from the brown alga Eckloniakurome on red tide microalgae. *Aquaculture*, 2003, 218, 601-611.

Onofrejova, L; Vasickova, J; Klejdus, B; Stratil, P; Misurcova, L; Kracmar, S; Kopecky, J. andVacek, J. Bioactive phenols in algae: the application of pressurized-liquid and solid-phase extraction techniques. *Journal of Pharmaceutical and Biomedical Analysis*, 2010, 51, 464-470.

Pietta P-G. Flavonoids as antioxidants. *Journal of Natural Products*, 2000, 63, 1035-1042.

Quideau, S ; Deffieux, D ; Douat-Casassus, C ; Pouysegu, L. Plant Polyphenols: Chemical Properties, Biological Activities, and Síntesis. *AngewanteChemie International Edition*, 2011, 50, 586-621.

Rossano, R; Uungaro, N; D'ambrosio, A; Liuzii, GM. and Riccio, P. Extracting and purifying R phycoerythrin from Mediterranean red algae *Corallinaelongata* Ellis and Solander. *Journal of Biotechnology*, 2003, 101, 289-/293.

Shibata, T; Fujimoto, K; Nagayama, K; Yamaguchi, K. and Nakamura, T. Inhibitory activity of brown algal phlorotannins against hyaluronidase. *International Journal of Food Science and Technology*, 2002, 37, 703–709.

Shibata, T; Kawaguchi, S; Hama, Y; Inagaki, M; Yamaguchi, K. and Nakamura, T. Local and chemical distribution of phlorotannins in brown algae. *Journal of Applied Phycology*, 2004, 16, 291-296.

Steinmetz, KA. and Potter, JD. Vegetable, fruit and cancer epidemiology. *Cancer Causes and Control*, 1996, 2, 325-351.

Southon, S. Increased fruit and vegetable consumption within the EU: Potential health benefits. *Food Research International*, 2000, 33, 211-217.

Targett, NM. and Arnold, TM. Predicting the effects of brown algal phlorotannins on marine herbivores in tropical and temperate oceans. *Journal of Phycology*, 1998, 36, 195-205

Targett, NM; Boettcher, AA; Targett, TE. and Vrolijk, NH. Tropical marine herbivore assimilation of phenolic-rich plants. *Oecologia*, 1995, 103, 170-179.

Yangthong, M; Hutadilok-Towatana, N. and Phromkunthong, W. Antioxidant activities of four edible seaweeds from the Southern Coast of Thailand. *Plant Foods Human Nutrition*, 2009, 64, 218–223.

Yu, S-S; Hu, Y-C; Wu, X-F. and Liu, J. Natural phloroglucinols-molecular diversity and bioactivity. *Recent Progress in Medicinal Plants*, 2009, 25, 103-139.

Yuan VY. and Walsh, NA. Antioxidant and proliferative activities of extracts from a variety of edible seaweeds. *Food and Chemical Toxicology*, 2006, 44, 1144-1150.

INDEX

C

D

E

F

G

H

I

J

K

L

M

N

O

P

Q

R

S

T

U

V

W

X

Y

Z